Paul Somerfield.

NORTH BULL ISLAND DUBLIN BAY

NORTH BULL ISLAND DUBLIN BAY
-a modern coastal natural history

Edited by
D. W. Jeffrey – General editor
R. N. Goodwillie
B. Healy
C. H. Holland
J. S. Jackson
J. J. Moore

Designer
Jan de Fouw

The Royal Dublin Society

First Published by The Royal Dublin Society 1977

ISBN 0-86027-001-7 casebound
ISBN 0-86027-002-5 paperbound

This book has been set in 12pt Bembo
and printed on Matt Coated Cartridge 115 gsm
Printed and Bound by Cahill (1977) Limited

CONTENTS

ACKNOWLEDGEMENTS

The editorial committee wishes to thank the following organisations and individuals who have made the production of this book possible at minimum cost to the purchaser. The Royal Dublin Society, through its Science Committee, gave encouragement at an early stage and has underwritten the major costs of publication.

As a practical contribution to conservation, sponsorship has made possible the inclusion of colour plates and has helped in production and publicity. Our thanks are due to:

I.C.I. (Ireland) Ltd.
Dublin Port & Docks Board
The Bank of Ireland
Tara Mines Ltd.
Asahi Chemical Industry (Ireland) Ltd.
An Taisce
Dublin Corporation
Dublin Regional Tourism Organisation Ltd.
Dublin Naturalists Field Club in memory of J. H. Brunker
Association of Geography Teachers of Ireland
Allied Irish Banks
Royal Dublin Golf Club
Arnotts Ltd.

The following generously made photographic material available without royalty: Richard T. Mills; Tom Keogh; Rex Roberts Studios Ltd.; J. P. Newell; J. Shackleton; Barry Kinsel; G. van Gelderen. Most of the maps are based on an aerial photo-mosaic produced by Fairey Surveys of Ireland Ltd. Cover photograph by P.D.I. Ltd., Dublin.

The secretarial work entailed in publication has largely been carried out in the organisations associated with authors and editors. This is a substantial contribution and we thank both the organisations and their secretaries.

Except where specifically acknowledged, illustrations are the work of authors and editors with additional help from Michael Telford.

PREFACE

This book is intended to be a means to the appreciation and study of a series of natural habitats within a city. It is not an encyclopaedia but rather a modern natural history produced by a co-ordinated team of specialists for the general reader. There are several ways of using it. Some readers may start at Chapter one, and work systematically through to Chapter eighteen. If they use the glossary, resort to further reading, and visit the island throughout the year they will have received a substantial course in coastal studies. The island is an exciting place at all seasons and is visibly changing from year to year. Other readers' needs may be more specialised: how do you explain to young children why the dunes lie beyond the beach; what features might be suitable for a Leaving Certificate project; just what should an administrator consider in devising management policy? Amateur naturalists will hopefully be well served, especially in broadening their interests and stimulating further study.

In a more specialised way still it contains an authoritative catalogue of organisms so far recorded. The check lists are tools for the naturalist, serving as a guide for identification, a basis for ecological studies, and a description of the biological *status quo* in 1976. Readers remote from Dublin can use most of the material in interpreting other dune, salt marsh, mud flat or beach situations. They will find that this book deals with a wider range of material than any other single publication dealing with estuaries or coasts.

The 'Bull Island Project' started informally late in 1974, and gained momentum by the establishment of an editorial committee by the Royal Dublin Society. This committee, with C. H. Holland, Dept. of Geology, Trinity College, as chairman sought contributions of written material and eventually financial support. Each member of the committee has edited a group of chapters, with the general editor being responsible for overall co-ordination.

A successful symposium, held at the Royal Dublin Society in December 1975 reflected the widespread interest generated by the project throughout Dublin. The reason for the enthusiasm is simply that the Bull Island is the most important educational resource for coastal studies in the country. It is a site where educational, scientific and recreational amenities overlap and co-exist. In producing the volume we have tapped a source of co-operation and goodwill which is the key to the survival of these values.

D. W. Jeffrey, General Editor.

Plate 1:1. Part of a map of Dublin Bay made by Captain William Bligh, R.N., in December 1800 and published by him in London, April 12th, 1803. The proposed wall on the north side of the estuary channel, as marked by Bligh, was not accepted. (From a map in the possession of John W. Griffith, Esq.).

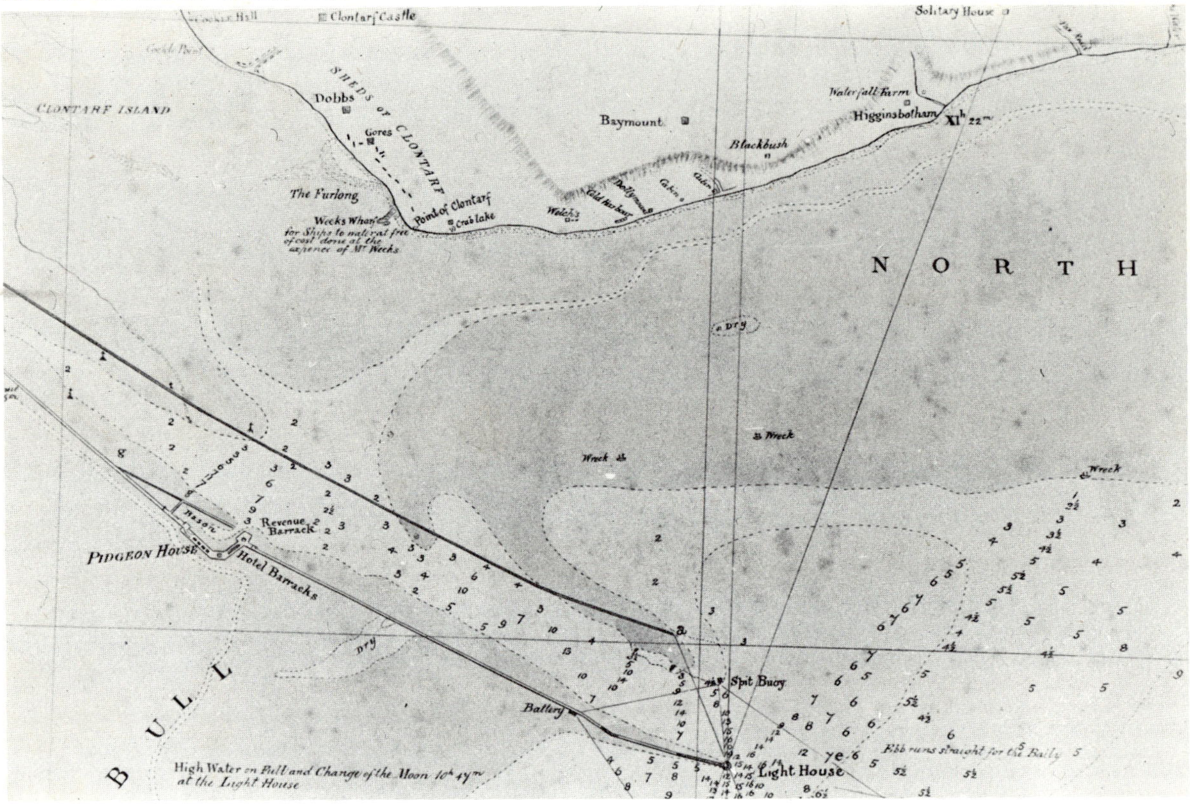

Plate 1:2. Map of the North Bull made by Francis Giles in 1819 to indicate the position for the new wall, and the area of the surrounding tidal ground to be acquired by the Ballast Board. (From a map in the possession of the Dublin Port and Docks Board).

Chapter 1

HISTORICAL EVIDENCE FOR THE GROWTH OF NORTH BULL ISLAND

Donal T. Flood

On 13th October 1801, Richard Broughton, as secretary to 'The corporation for improving and preserving the Port of Dublin', commonly known as the Ballast Board, in a letter addressed to the Directors-General of Inland Navigation in Ireland, commented that 'the north bull . . . has increased very much, both in extent and height, during the progress of the works on the south side, and more especially since their completion'. The Directors-General had been given responsibility for the improvement of Dublin Harbour by the Irish Parliament in 1799, and it was for that purpose that Broughton advocated 'running a work from the dry part of the bull to the shore'. The previous January, Captain William Bligh, arising from a survey of Dublin Bay, which they had commissioned, had presented the Directors-General with a map *(Plate 1:1)*, on which the dry part of the North Bull was marked as a very small patch. It was immediately off-shore from the large 18th century mansion known as Baymount (now Manresa). In their own submission to the Lord Lieutenant, Cornwallis, in 1804, the Directors-General supported the proposal of the Ballast Board and referred to 'the degree of accumulation which has already appeared on the north bull since the building of the south pier, so great that a considerable stripe of it remains dry at high water and has on it a growth of marine plants'. The small oval patch, which Bligh had mapped, in three years had become 'a considerable stripe'.

It was not until 1814 that the Ballast Board, by the sale of a site at the Pigeon House for the erection of a fort, secured sufficient money to proceed with their proposal of 1801. In 1818, they engaged an English engineer, Francis Giles, to prepare a plan *(Plate 1:2)* for a wall to extend from the Clontarf shore to the North Spit of the Pool Beg. On the map, which Giles prepared in 1819, the dry patch was depicted as having grown into a spear-shaped island about 3 km long. While he marked it on that map as 'The Sand Island', Giles referred to it later as 'the Green or Bull Island'.

In collaboration with George Halpin, Inspector of Works to the Ballast Board, Giles determined the site for the new wall on a more direct line to the Spit Buoy than that originally proposed in 1801. The timber bridge, which represented a fundamental change in the original plan, was built under separate contract in the autumn of 1819. In his letter of 1801, Broughton had indicated that there was a problem concerning the cutting off 'the gut eastward of Clontarf', that is the channel north of the Bull. Giles made a strong case for leaving this channel open by providing a gap 600 ft (180 m) wide at the shore end of the wall. The wall itself was completed by 1825. The purpose was 'for making the Harbour a secure Station for the Trade and improving the approach to it'. While the wall provided a barrier to the encroachment of sand on the harbour, it did not otherwise affect the growth of the island.

The Great North Wall, as it was officially designated, or the Bull Wall, as it became popularly known, had first been proposed by an English engineer, William Chapman in 1786. His proposal later found support from Captain Daniel Corneille, who was engineer to the Ballast Board. At the turn of the century, attention had become concentrated on the navigational hazard presented by the sand bar at the entrance to the harbour, following the construction of the Great South Wall.

For many centuries, during which Dublin had become the principal importing centre for the country, the shallowness of the approaches left the quays of the old city inaccessible to deep-sea ships. Trade was dependent on a deep-sea anchorage in Dalkey Sound, and on shallow anchorages in the Pool Beg, Clontarf Pool, and elsewhere in the open estuary. Cargoes were trans-shipped there into sailing barges (gabbards) or smaller boats, for unloading at slipways which reached from the city quays into the river channel. As trade increased, and, with it, the demand for space on the city quays, it became necessary to unload some vessels at low tide, on the Green Patch, or in the Salmon Pool, and carry the merchandise on litters across the sand to the roadway at Ring's End, for carting to the city. By the end of the 17th century, the Dublin merchants, faced with increasing losses of ships damaged, while at anchor, by pounding against the hard bottom at low tide in stormy weather, petitioned the Irish Parliament. They secured legislation in 1707 by which authority was given to Dublin Corporation to control the digging of ballast in the river channel and to straighten and deepen the approach by water to the city. This authority was transferred to the Ballast Board when it was established in 1786.

The great engineering achievement, by which the port of Dublin was created in the 18th century, was made possible by the presence of gravel and boulder clay, which not only provided a firm foundation for the Georgian houses with their deep-dug basements but extended well into the Bay. Commencing at Essex Bridge (Parliament Street), in 1711, a new and deeper channel, embanked on both sides, was constructed as far as the outlet of the Dodder above Ringsend. The embankment on the south side of the channel was continued across the South Bull, for about 5 km, to the eastern limit of low water. The construction was effected by driving timber piles deep through the over-lying sand and reinforcing them with baskets (kishes) filled with gravel. Much of the gravel was drawn by boat from Clontarf; Clontarf Head and the greater part of Clontarf Island, as well as Ring's End, which were featured on 17th century maps, had disappeared when Bligh drew his map in 1800. Much of the remains of Clontarf Island provided gravel for Bindon Blood Stoney's pioneering use of concrete in the mid-nineteenth century construction of the Alexandra basin.

The embankment across the South Bull, which was known as 'The Piles', was completed about 1730. It was successful in providing shelter for the various anchorages in the estuary, and in preventing the drift of sand into the channel, but it was eventually breached by a storm when the timbers rotted. In 1761 it was decided to replace it by a stone pier. Commencing at the seaward end, the present lighthouse at the Poolbeg was completed in 1768, but the present South Wall, built of massive granite blocks, did not reach the Pigeon House Harbour until 1796, after very slow progress.

These were the 'works on the south side' to which Broughton had referred in 1801. Concurrent with their construction, gravel and sand were dredged from the channel at the rate of 100,000 tons each year. Some part was sold as ballast, but much was used as infill for the new walls, as well as the North and South Lotts, and for the old river channel, on which a new Custom House had been built during the years 1781–1791. At the close of the century, sailing ships of 300 tons burden could find sufficient depth to anchor at the Custom House. Dredging continued and it has been estimated that more than 32 million tons were removed from the harbour-bed during the nineteenth century.

Meantime, the Bull Island continued to develop. As Broughton had suggested, the increase in level was attributable to the construction of the South Wall. Initially, it was assumed that the sand-banks between the Pool Beg and the Dodder had merely been moved across the harbour to the east by the tides. Indeed, in their submission of 1804, the Directors-General devoted considerable argument to the movements of sand within the Bay, to support their thesis that the building of the Bull Wall would concentrate the flow of the ebbing tide and shift the sand-bar which lay across the harbour entrance *(Plate 1:3)*. This bar restricted entry during the six-hour period when the sea was below the half-tide mark. But they failed to recognise that fresh sand was carried into the Bay by every turbulent tide rising across the sandbanks off the Wicklow and Wexford coast.

However a local movement leading to the build-up on the North Bull was recognised. By an important experiment which Captain Corneille made, at the request of the Directors-General, to ascertain the drift of the sand on the bar, some observations which are fundamental to explaining the growth of the island were recorded. A large raft, some five metres square, had been built from heavy timbers and was moored over the sand-bar. Long poles extended from it to reach the sandy bottom, and these were free to move up or down as the raft rose or fell. With the movement of the raft by the waves, the sand on the bottom was loosened and disturbed. Corneille reported: 'it was observed that in various winds the first of the flood carried the sand towards Sutton; that in the progress of the flood-tide the sand was carried northward

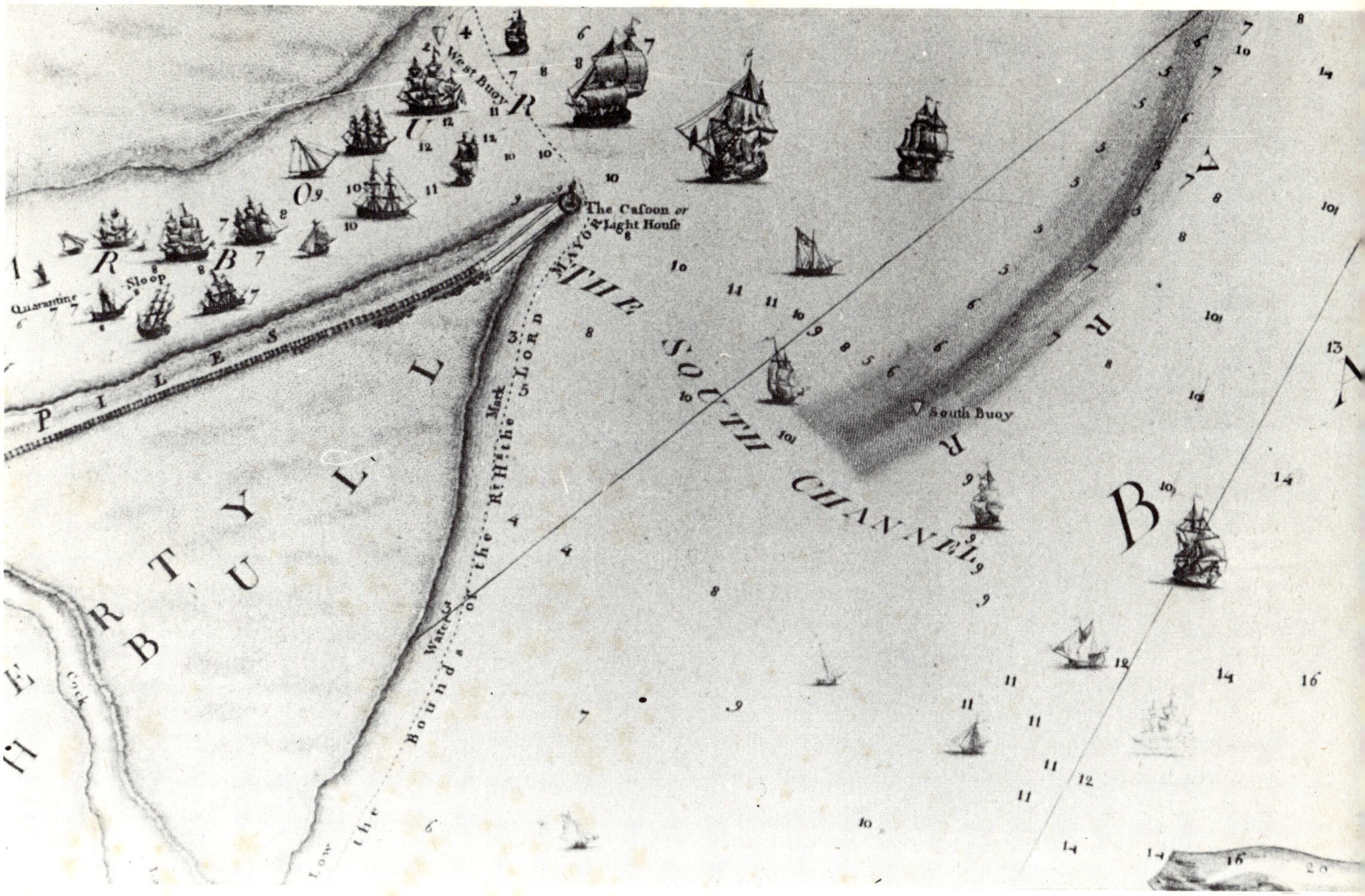

Plate 1:3. Part of a map of Dublin Harbour made by Bernard Scale in 1773. It shows 'The Piles'; the Poolbeg Lighthouse, completed in 1768; and the early construction of the present South Wall, together with the 'Barr'. (From a map in the possession of John W. Griffith Esq.)

directly upon the north bull, and lastly in a north-west direction towards Clontarf'. This tidal movement, is, of course, a result of the main flow northwards of the rising tide in the Irish Sea, and it can be related to the tidal movement in the channel north of the Bull Island.

Corneille's observations of the tidal currents within the northern part of the Bay provide an explanation of the detachment of the North Bull from the north shore. In an undated report to the Directors-General, written between 1802 and 1804, he wrote: 'At present a very great proportion of the tide flows into the harbour by Sutton Creek, round the northward of the north bull along the Clontarf shore, and its influence is even seen so far up in the harbour as the back of Clontarf Island. . . . Now upon the return of the tide or ebb at present a great body of water . . . escapes or runs out through the channel at the back or northward of the north bull by Sutton Creek, from the time of high water until the half-ebb, after which time its passage is prevented by the height of the sand in this channel, opposite to or near Cold Harbour'.

The tidal current through the channel was responsible also for creating the slob or mud flat which was the forerunner of the salt marsh on the north side of the island. For many centuries much of the waste of the city, including the sweepings from the streets, was consigned to the Liffey for removal by the ebbing tide. The southward-flowing ebb carried it towards Killiney where it was met by the rising tide and carried across the Bay towards Howth. The incoming tide then carried it up Sutton Creek and into the north channel. The Directors-General were at pains to prove that none of the material carried by the river found its way onto the bar: 'The stuff which the river brings down in common weather is no more than soft ooze from the sewers of the city; and what is not sufficiently heavy to subside between the walls, is floated all over the harbour, and deposited on or near the shores. . . . The fine particles of ooze or loam which float long in the water are dispersed, and such parts of them as are light enough to reach the salt water . . . will not sink upon the bar, but be carried to the higher shores. Accordingly, no such deposit of ooze or soft stuff is found upon the bar; the sand there is clean as on either of the bulls; and the soft ooze is found . . . on the strand between Clontarf and Ballybough, and the northern parts of Sutton Creek'.

The deposition of ooze must have continued through the nineteenth century, at an ever increasing rate, as the city grew and water-closets came into wider use. The present main drainage system, which intercepted the old city sewers, for treatment of the sewage at the Pigeon House Harbour, was not completed until 1906. Evidence that a slob still existed at that time has been afforded by two photographs published by C. J. Patten. One, which is identifiable as of the entrance to Sutton Creek, bears the caption: 'The Bare Slob-Lands at First Ebb'; the other is identifiable from its background as of the north-west end of the present marsh. It bears the caption: 'An Estuarine Mud-flat': the elevated patches of soft ooze, covered with dark-green slimy seaweeds, are here and there intersected by salt-water channels and pools . . .'. It was in this area, also, that E. A. Letts in 1907 found that plants of *Ulva latissima,* bearing fronds over three feet (about one metre) long and one to two feet broad, occurred frequently. He had demonstrated experimentally the capacity of this alga, commonly called sea-lettuce, and also that of the brighter green *Enteromorpha,* to absorb rapidly both ammonium salts and nitrates, the end-products of the bacterial decomposition of organic nitrogenous matter.

Patten's second photograph, demonstrates also the erosion which is slowly destroying the marsh. The area shown has now lost its elevated patches and is covered by every tide; and *Ulva* has been displaced by *Enteromorpha.* The large area of the saltmarsh itself has lost its earlier organic content and now supports a different type of vegetation. Paved streets, free from the horse-dung of other days, and a sewage plant, to separate the solids, have altered the nature of the effluent which still affects the estuary. But the cockles and mussels, which Molly Malone cried 'in Dublin's fair city' of the eighteenth century ballad, remain evident in the huge quantities of shells on the harbour-bed and by the shells cast upon the beach of the island.

The first sand dune of 1800, from which the supratidal island has grown, doubtless developed on the edge of the slob which had formed from the city's 'ooze' on the highest part of the North Bull sands during previous centuries. The Admiralty chart of 1838 indicates that the highest point on the island had reached 24 ft (7.3 m) 'above low water', or about eleven feet (3.4 m) above high water.

In 1868, a serious threat to the future of the island followed the granting of a concession by the Corporation to Barrington and Jeffers to dispose of the city sewage. The City Engineer, Parke Neville, referred to the project in his report of 1869: 'The promoters propose to utilise the sewage partly for the purpose of reclaiming portions of the sands and sloblands of the North Bull'. The opposition of the coastal landlords, Benjamin Lee Guiness of St Anne's, with the Vernons and Lord Howth, defeated the project. The President of the Institution of Civil Engineers, J. F. Bateman, referred to this proposal in 1879, when speaking to a paper presented by John Purser Griffith. As consultant to Barrington and Jeffers, he had found that between 1819 and 1865 the depth of sand at the Bull Wall had increased by eight feet (2.4 m): a half-mile (0.8 km) farther along the island it had increased by eight and one half feet (2.6 m), and the margin of low water had been extended upwards of a half-mile farther south. Farther along the island, the amount of the increase in level became less, and at a distance two and one half miles (4 km) from the wall, the level of the sand and the margin of low water remained the same as Giles had recorded in 1819.

Chapter 2

SEDIMENTOLOGY AND GEOMORPHOLOGY

C. R. Harris *Department of Geology Trinity College*

North Bull Island is a low-lying, dune-covered sand spit. It is 4.85 km long and runs parallel to the coast in a north-easterly direction between Clontarf and Sutton in the northern half of Dublin Bay. At low tide, extensive sand flats occupy the inner third of Dublin Bay. These form the North and South Bulls, better known as Dollymount and Sandymount respectively. They are separated by the combined estuaries of the Liffey and Tolka rivers. The name 'Bull' may be a derivation of the term 'ball' meaning an offshore sand-bank. The supra-tidal part of the island is the backbone of dune ridges. Landward of this is a salt marsh that encroaches upon a lagoon. This drains at low tide exposing sand and mud flats. Two permanent channels exist in the lagoon, one carrying the Naniken Stream south-westwards under Bull Bridge and the other carrying the Santry River north-eastwards to Sutton Creek. The island is terminated to the south-west by the Bull Wall beyond which, and projecting into the Tolka estuary, is the intertidal North Spit sand-bank. This is separated from the Clontarf shore by Crablake Water. To the north-east the island tapers off and is separated from Howth Head by the deep Sutton Creek.

North Bull Island is a favourite recreational area for the people of Dublin, because of its proximity to the city and its extensive beach. This is accessible to motor traffic by a narrow wooden bridge, Bull Bridge, connecting Bull Wall to the mainland, and by a more recently constructed causeway extending across the lagoon and island from Watermill Lane. This causeway has divided the lagoon into a northern and a southern section.

A group of cottages was built on the island next to the Bull Bridge as coastguards' dwellings. The only other buildings, apart from bathing shelters and lifeguards' huts, are the club-houses of two golf clubs whose courses occupy most of the dune area.

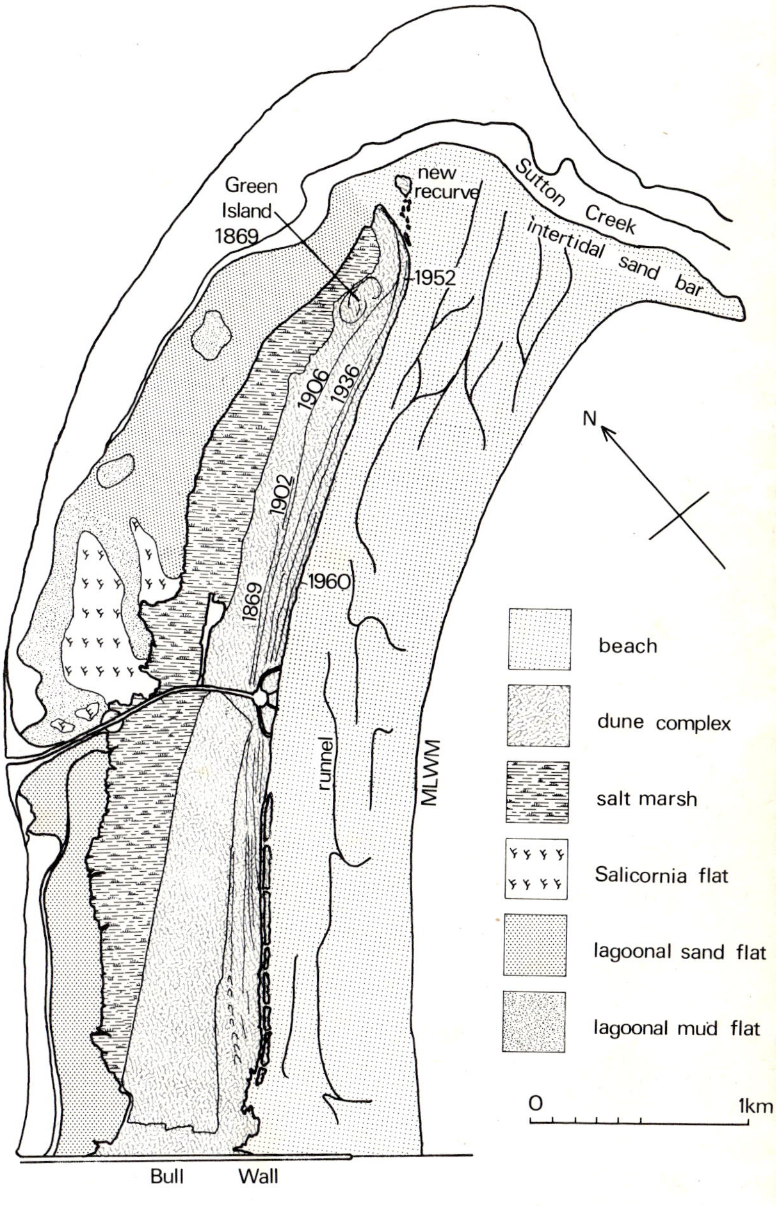

Fig. 2:1. Sedimentological subenvironments of North Bull Island.

North Bull Island – Dublin Bay

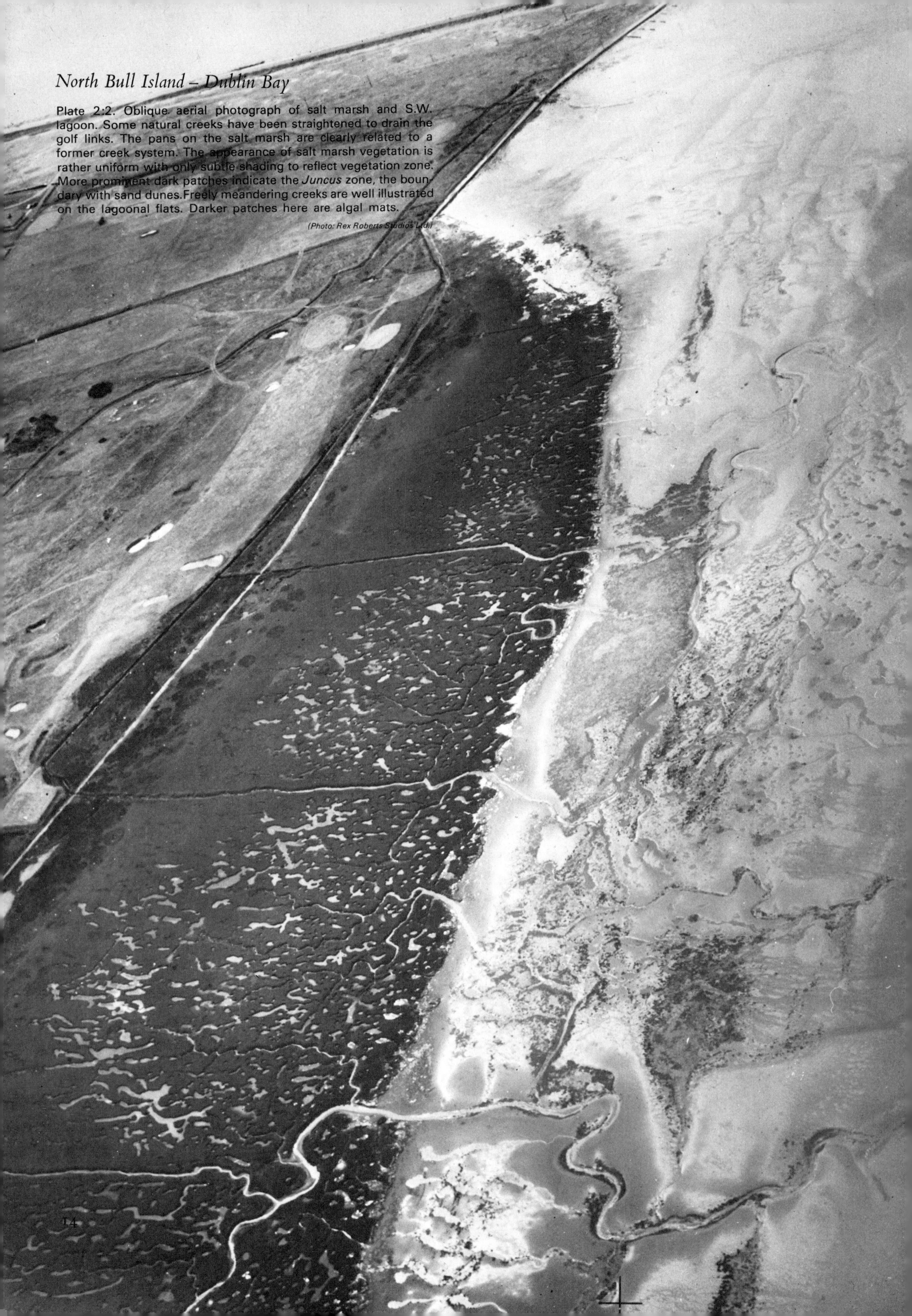

Plate 2:2. Oblique aerial photograph of salt marsh and S.W. lagoon. Some natural creeks have been straightened to drain the golf links. The pans on the salt marsh are clearly related to a former creek system. The appearance of salt marsh vegetation is rather uniform with only subtle shading to reflect vegetation zone. More prominent dark patches indicate the *Juncus* zone, the boundary with sand dunes. Freely meandering creeks are well illustrated on the lagoonal flats. Darker patches here are algal mats.

(Photo: Rex Roberts Studios Ltd.)

Geological origins of the island

During the Pleistocene period ice extended southwards along the present site of the Irish Sea, ending in terminal moraines between Carnsore Point and St David's. Ten thousand years ago, at the close of the Pleistocene, sea-level was some 30 m lower than at present and the exposed part of the floor of the Irish Sea was blanketed by glacial and fluvio-glacial debris. During the ensuing Flandrian period, sea-level rose and dominant southerly currents sorted and re-distributed the Pleistocene sediment. The finest particles were transported to the northern areas of the Irish Sea while around Dublin Bay predominantly sandy sediment accumulated. In Dublin Bay about five thousand years ago sea-level had risen to a maximum height of 4 m above the present level, forming the raised beach evident along the Kilbarrack shore and the tombolo connecting Howth Head to the mainland.

North Bull Island has appeared within the last two hundred years after improvements (referred to in the previous chapter) were made to the Port of Dublin. The building of the South and Bull Walls altered the configuration of currents in the Bay and an island appeared on the sand-banks of the North Bull *(Plate 1:2)*. However, there was a system of ridges and runnels on the sand-banks long before the island appeared. This latter event occurred with accumulation on and subsequent emergence of the most landward ridge.

Due to the prevailing current direction, the North Bull has developed in the part of Dublin Bay that acts as a sediment trap. Thus, as more and more sediment accumulated in the area, the intertidal ridge built up towards the high tide level. On reaching this critical height further elevation occurred only by wind-blown sediment and, eventually, grass-stabilised dunes formed.

Once a supra-tidal sand-bank has been established, its growth direction is determined by the combination of current direction, wave direction, wave energy, the amount of sediment available, and water depth. In the case of Bull Island, growth direction was parallel to the shore, both south-westwards and north-eastwards, because this was the orientation of the intertidal ridge along the crest of which emergence had first occurred.

A fundamental factor in the sub-aerial elongation of a sand-spit is the presence of an intertidal spit-platform on which it can grow. Where the platform is close to the high tide level, elongation will proceed more rapidly than where the platform lies in deeper water. The height of the intertidal spit-platform decreased along its length to the south-west as it fell away into the Liffey-Tolka estuary, and to the north-east as it approached Sutton Creek. After extension in both directions, south-westerly growth was halted by the construction of the Bull Wall but growth continued unimpeded towards Sutton Creek assisted by dominant wave-fronts flowing from the south-south-east.

Hydrography

Gross tidal movement in Dublin Bay is clockwise. However, tidal flow affecting the foreshore of Bull Island is more complex, due to interference by the Bull Wall, the Liffey – Tolka outflow, and the flow through Sutton Creek. Flood tide movement on the beach is longshore to the north-east. During the ebb-tide, this direction is maintained but farther offshore the water is caught up in the substantial fluvial 'jet' which emanates from between the Bull Wall and the South Wall and flows around Howth Head.

The lagoons landward of Bull Island are filled and emptied through two channels. In the southern lagoon the flood tide enters beneath Bull Bridge. It fills the permanent channel, which also acts as an outlet for the Naniken Stream, and inundates the sand flats. As the tide falls there is a sheet flow of water off the sand flats into the channel which conducts the water back under Bull Bridge. In the northern lagoon water is channeled in and out through Sutton Creek. Again, a permanent tidal channel drains the lagoon, also carrying water from the Santry River.

Geomorphology of the subenvironments

The surrounding intertidal flats are a complementary and integral feature of Bull Island, which can thus be divided into two main environments, supra-tidal and intertidal. For precise descriptive purposes this division is too broad and six sub-environments have been erected on the basis of surface morphology and sedimentological character *(Fig. 2:1)*.

The beach

Bull Island strand runs the whole length of the island and, at low water, its width varies from 880 m at the Bull Wall *(Plate II)* to 620 m seaward of the new St Ann's Club House. At the Sutton end an intertidal sand-bar stretches out 1375 m from the dunes into the Bay alongside Sutton Creek. The beach may be divided into upper and lower parts, the backshore and foreshore respectively. The line of separation is the berm, recognisable as a change in slope formed by material accreted at the high-tide level. This change in slope is further emphasised by an accumulation of shells, weed, detritus, and, locally, vast numbers of tubes of the sand mason, *Lanice conchilega*.

The most prominent feature of the beach is the system of ridges and runnels aligned parallel to its length. The ridges are asymmetrical in cross-section, the steep slope facing landward and the gentler slope facing seaward. In general, the top and landward sides are rippled while the lee-sides are smooth. The ridges and runnels increase in number as the beach gradient becomes less steep. Thus, only two occur at the Bull Wall while six ridges occur along the sand-bar at the north-eastern end of the island. Interconnecting drainage channels run either normal or oblique to the trend of ridges and runnels, carrying away the water stranded in the runnels during ebb-tide and filling them again with the rising tide. Even at low water the runnels are not completely evacuated, for a certain amount of water is locked in by the slight elevation of the runnel bed where it is deflected as a channel through the ridge.

The ridges are a type of swash-bar; that is they form directly under the break-point of a wave, where water velocity is suddenly checked and sediment is thus deposited. They have developed on

Plate 2:1. View of the 1936 recurve, looking north-eastwards from the salt marsh.

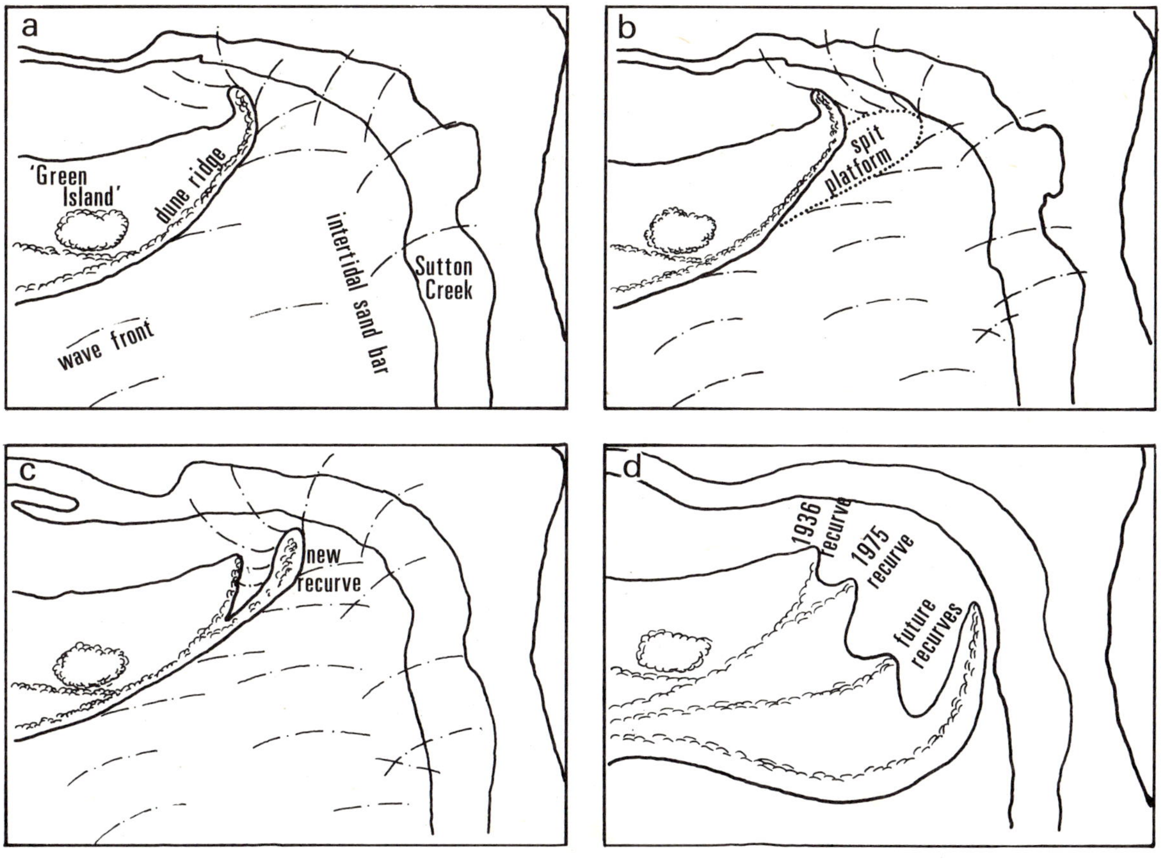

Fig. 2:2. **(a)** The situation in 1936. Wave-fronts fashion a typical recurved spit. **(b)** Spit-platform becomes sufficiently elevated to re-configure wave-fronts, which start to comb-down the recurve tip. **(c)** The situation in 1975. A new recurve grows on the spit-platform, while destruction of the old recurve continues. **(d)** Projected development of recurves along the intertidal sand-bar.

North Bull Island through a steady, abundant supply of sand by offshore currents which has produced a wide beach with a low basic gradient which varies from 1:150 to 1:350. On the other hand, the restricted fetch of the Irish Sea produces short waves which have the potential for building beaches with steep gradients. In the resulting gradient disequilibrium ridges occur where the waves locally impose their swash gradient upon the breach profile.

More ephemeral features, such as ripples and other bedforms, present on the beach will be described in the section dealing with sedimentary structures.

The dune complex

This sub-environment forms the island's backbone. It is 880 m wide at the Bull Wall and tapers off in the north-east to a recurved hook. At this point the dunes attain their maximum height of 10.21 m above Irish Ordnance Datum. Elsewhere they average a height of 6 – 7 m. Much of the dune complex has been modified by the golf-links, although dune ridges are still visible.

Several distinct types of dune can be recognised. The most prominent are the virtually parallel dune ridges *(Plate IV)* separated by more densely vegetated valleys or slacks. Shorewards of these is an irregular arrangement of lower (one metre high) incipient dunes, or foredunes. These are backshore deposits newly stabilised by colonising grasses. At the north-easterly tip of the island they can clearly be seen assisting in the process of developing a new recurve. Parabolic dunes occur north-east of the Royal Dublin Club House. These open to the south-west and have been formed by strong winds from that direction, blowing out parts of the existing dune ridges.

Towards the tip of the island, two interesting features can be seen. The first is the Alder Marsh, a wide depression caused by the divergence and later convergence of two dune ridges. The second is a patch of circular dunes just beyond it. An understanding of the origin of these two phenomena lies in the study of old maps of the island. By comparing the position of dune ridges on aerial photographs with Ordnance Survey maps showing the island's changing shorelines, from the first publication in 1837 to the latest in 1936, a distinct correlation can be seen. It is possible, therefore, to date the ridges and relate them to ancient shorelines of the island. Some of these are shown in *Figure 2:1*.

With regard to the Alder Marsh, the ridge bounding it on the landward side corresponds to the 1906 shoreline, while the one to seaward represents the shoreline as depicted on the 1936 map. The Alder Marsh surface is topographically the lowest point *(Fig. 3:3)* in the dune complex, and thus is closest to the fresh water table. Water logging in winter and an ample supply of water in summer have led to the accumulation of nearly 30 cm of peat-like organic matter above the sand.

Using old maps and modern aerial photographs, the origin of the circular dunes also can be clarified. These are located at the site of a 'Green Island' as shown on the 1869 O.S. map. At that time, the growing Bull Island had not extended so far and an isolated supra-tidal bank was in existence there. From the name of the bank, it was presumably composed of grass-covered dunes. Subsequent growth of Bull Island enveloped Green Island by 1902.

Salt marsh

This sub-environment occurs between the dunes and the lagoon. Its boundary with the dunes is marked by limitation of penetration of highest spring tides and by a sharp vegetational change. In places 'wedging' of blown sand does make the boundary more complex. On the lagoonal side there is a distinct boundary in the form of a 'cliff', up to 40 cm high, *(Plate I)*, aproned in some places by a zone of mud-balls formed by disintegration of the marsh *(Plates 2:1 & V)*. This 'cliff' coincides with the normal high-water limit, and only the spring high-tide rises above this level, flooding the marsh. Stratigraphically, the marsh is a relatively thin crust of sandy mud with much plant debris which is spreading out over the lagoon from the dunes. It is thickest along the lagoonward edge but, even so, nowhere exceeds a depth of 30 cm.

Two morphological features may be seen on the marsh. The first is a series of natural drainage creeks *(Plate XII)*. These meander generally north-westwards into the lagoons. They have cut right through the salt marsh crust and into the underlying lagoonal sand. Slight elevations of the creek beds at the marsh edge prevent them from draining completely. The reduced salinity of the creek water at low tide may be due to additions of rain water, surface runoff, and even, if the creek is deep enough, of fresh water tapped from the non-saline ground-water table. Also, scattered across the marsh surface are pans generally a few meters in diameter *(Plate 2:2)*. These are shallow (10 cm), steep-sided, sub-circular depressions floored with bare mud. The majority of pans were probably associated with creek systems which have been dammed up and broadened by wave action during the course of time.

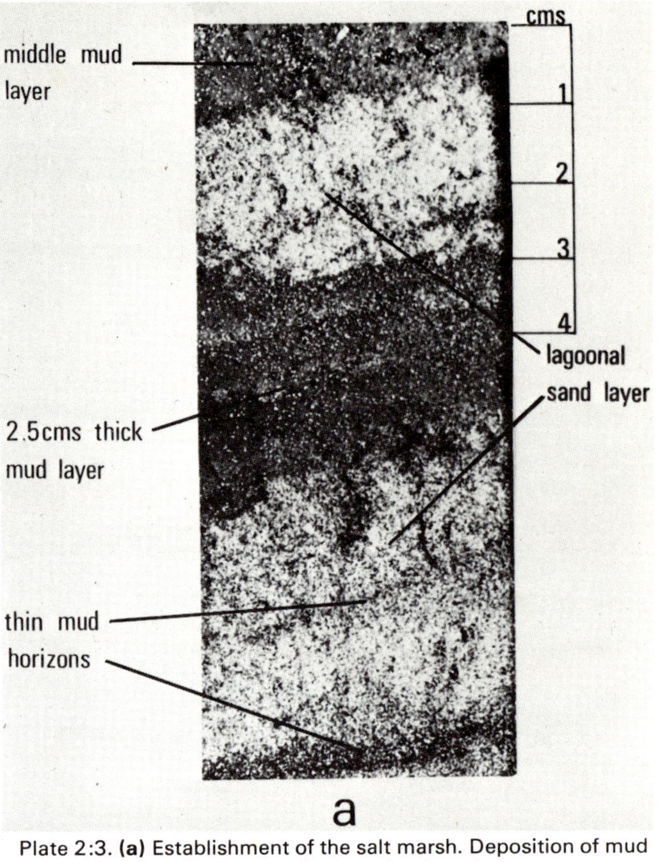

Plate 2:3. **(a)** Establishment of the salt marsh. Deposition of mud on lagoonal sand flats occurs firstly as laminae, later as more persistent mud layers **(b)** Sandy horizon in upper peat layer caused by sand being deposited upon the marsh surface by storms. **(c)** Thin layers of sand and mud irregularly deposited by current action near the 1936 recurve. [Harris].

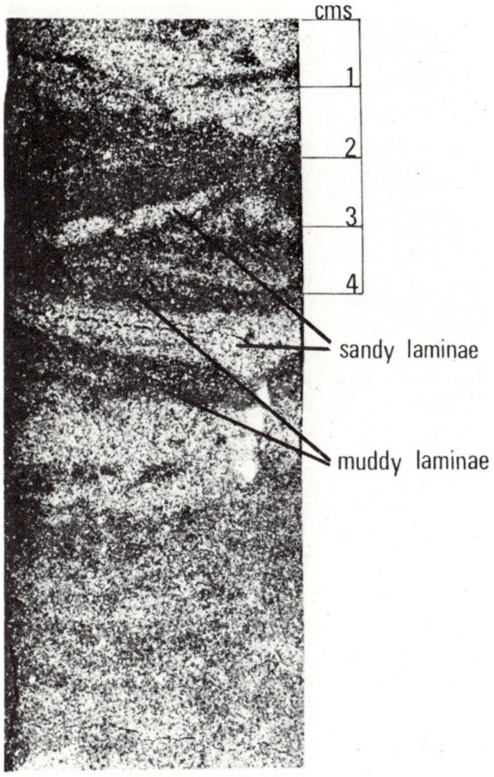

c

b

The way in which the marsh spreads out across the lagoon can be understood by examining vertical cores of marsh sediment. There are three layers in the marsh. The lowest layer is slightly muddy sand comparable with that which forms the lagoonal sand flats. Above this a middle mud layer occurs which, near the surface, becomes an upper peat layer through ramifications of marsh plant roots. This suggests the deposition of mud associated with colonisation by marsh plants which assist in its accumulation. Perusal of the marsh cores shows that, in places, the depositional history is more complex.

For instance, in the core illustrated in *Plate 2:3 a*, the lowest layer shows mud forming thin horizons in the sand. This can be interpreted as marsh mud spread on the lagoon sand flat, but neither in sufficient amounts nor with sufficient continuity to establish a mud flat there. Even when, in this core, a mud layer 2.5 cm thick has formed, more lagoonal sand has been deposited upon it, either by storm driven high-tides or by mud starvation resulting in a return to sand deposition. Eventually, however, the middle mud bed became established.

In the core shown in *Plate 2:3 b*, even in the plant colonised upper peat layer, an influx of sand across the salt-marsh has produced a 1 – 2 cm thick sand layer. This feature occurs in several cores, suggesting that particularly violent storm conditions cause sand-drifts across the upper parts of the marsh from the dunes.

The characteristic sedimentary sequence of the salt marsh can also be demonstrated laterally at the marsh edge. Here the true lagoonal sands become buried by mud lumps eroded from the marsh. These form the middle mud layer. When this has built up sufficiently, marsh plants colonise it forming the uppermost, peaty layer. Marsh growth appears to be a periodic phenomenon. Erosion of the marsh cliff takes place forming the mud-ball apron *(Plate V)* and this is followed by further deposition on the apron and subsequent colonisation of it.

Lagoonal mud flat

The mud flat is largely restricted to that part of the northern lagoon adjacent to the causeway. Here they are typical 'schlick' deposits, thixotropic muds incapable of bearing a person's weight. Farther away from the causeway, they become firmer, muddy sand *(Plate VIII)* and finally grade into the lagoonal sand flat. The mud flat has been deposited within the last decade and overlays normal lagoon sand. Its thickness is not great – a little over 5 cm against the causeway and wedging out north-eastwards.

Its origin can be ascribed to the presence of the causeway and the radical hydrographical changes that it initiated in the lagoon. Before 1964, when construction was started, the flooding tide entered the lagoon under Bull Bridge and swept north-eastwards before ebbing through Sutton Creek. Little mud was deposited as currents were sufficient to prevent settling out of the suspended material. Since construction began the northern lagoon is filled by the flood tide entering through Sutton Creek. This water contains finely-particulate material in suspension which has been discharged from the Liffey. While some of this material floors the Bay off Dollymount, some is carried around Howth Head by the fluvial 'jet-stream' from the Port of Dublin and some is carried into the northern lagoon. At high water slack this material settles out in the quietest part of the lagoon, which is against the causeway. The ebb-flow elsewhere is sufficient to wash the suspended material back into the Bay. Two minor local sources for the mud can also be suggested. Material can be transferred from the salt-marsh sediments and the biodegradation of algae on the sand flats can produce organogenic mud. Furthermore, mud from the Santry River would find its way onto the lagoonal flat.

Two other, smaller patches of schlick occur in the northern lagoon. One of these is coincident with a large mussel-bank. The presence of this bivalve colony not only reduces water velocity in the area, thus causing deposition of fine-grained material, but also produces excremental muds. The second schlick patch occurs in a slight depression of the sand flat caused by a levee-like ridge of sand bordering the permanent channel which checks stream-flow across the lagoon into the channel, bringing about the accumulation of fine sediment.

The lack of current over the schlick patches has resulted in a generally smooth, unrippled surface.

Salicornia flat

This is a rapidly spreading sub-environment related to the largest schlick patch in the northern lagoon. It is characterised by the presence in summer and autumn of dense stands of the succulent annual halophyte *Salicornia (Plate XI)*. The stems of these plants serve to encourage the local accumulation of sediment. This leads to elevation of surfaces ultimately to that of the salt marsh level. Before construction of the causeway small stands of *Salicornia* occurred against the salt marsh, where they were protected from the main tidal flow and from the prevailing winds. After construction, parts of the lagoon adjacent to the causeway were sufficiently protected from tidal flushing to allow the spread of *Salicornia* over the mud and sand flats. While sedimentologically muddier, the *Salicornia* flat offers a firmer footing because it stands on slight elevations which provide better drainage for the sediment.

Lagoonal sand flat

The largest part of both lagoons is floored by sands having a small admixture of silt and clay *(Plate VI)*. A local source for this mud is the biodegradation of the alga *Enteromorpha*. The surface of the sub-environment has a corrugated appearance because the creeks, which emerge from the salt marsh, flow in a linear fashion across the sand flat towards the permanent channel. The sand flat is generally unrippled, except for a discontinuous narrow strip adjacent to the salt marsh edge where small, straight-crested ripples occur, and for the creeks, in which a disordered arrangement of linguoid ripples has developed.

Geomorphology of the recurve

The recurved hooks at the Sutton end of the island, typical features of growing sand-bars, are an excellent example of dynamic geomorphology *(Fig. 2:2)*. From cartographical data there is evidence that the first recurve developed on Bull Island around 1902. Until that time the supra-tidal part of the island had been growing steadily upon its spit-platform. This growth continued unimpeded until 1902 when the island had almost reached its maximum length. By then, the sand being made available for supra-tidal growth of the spit was deployed for the elevation of the platform, which sloped off

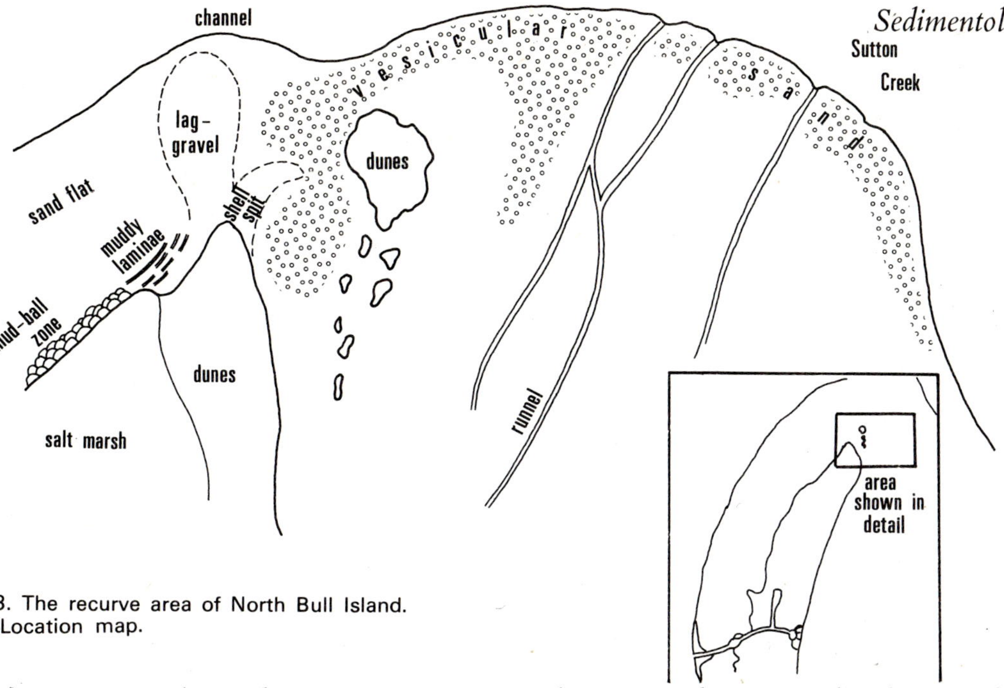

Fig. 2:3. The recurve area of North Bull Island.
Inset—Location map.

towards Sutton Creek. At the same time, a recurved hook began to form sub-aerially as the wave-fronts became refracted around the stationery tip of the island. A single set of refracting waves alone is sufficient to develop recurves. In this case, however, they were assisted by a second set flowing up Sutton Creek. Where mutual interference of these two wave sets occurred, a slight checking of current caused accumulation of sediment. This has been instrumental in the development of the bar growing alongside Sutton Creek, and also in the elevation of the platform to the north-east of the recurve.

Another feature of recurves is that, especially on well-established sand-spits, they do not occur singly. The process of recurve addition is a cyclic one of growth and destruction. Whilst attaining a classic recurve geometry, the spit-platform to the convex side of the hook becomes elevated. This, in turn, affects the wave-front configuration by destroying the waves before they reach the recurve and results eventually in the emergence of a supra-tidal bank beyond the recurve. By the time the waves have rounded this new projection, their incidence upon the recurve tip is destructive rather than constructive Thus, as one recurve declines another emerges.

According to dune-ridge patterns on Bull Island, no recurves existed before that of 1902. On the 1904 Ordnance Survey map, the recurve had suffered considerable erosion, due possibly to the seaward growth of the island. By 1936, in some 34 years, this recurve had re-established itself. Study of aerial photographs taken in 1971 and field observation *(Plate III)* showed that a new recurve had developed to the north-east in the intervening 35 years, the old one having undergone significant erosion.

The present distribution of sediments in the recurve area is shown in *Figure 2:3*. The recurve developed on shelly, supra-tidal deposits accumulated by storm-waves on the spit-platform. Upon these shell deposits wind blown piles of sand developed, some of which were stabilised by dune plants. Further sand accumulation thus was induced by entrapment, resulting in the growth of dunes and the establishment of the new recurve. Simultaneously, the reconfigured currents ate away the tip of the old recurve. Its previous extent can be seen at low water, for, stretching towards the Kilbarrack shore is a lag gravel of shells and pebbles *(Plate 2:1)*. Erosion is also taking place in the embayment leeward of the old recurve. Here, refracting waves are cutting into the salt marsh and a few metres from the present marsh edge, muddy laminae representing old salt marsh deposits are being exposed.

Although only three recurves have been recognised to date, it may be suggested tentatively that the periodicity of recurve growth-destruction on Bull Island is in the region of 35 years. Whether this stays true for future recurves remains to be seen. That other recurves will develop, however, appears likely. There is no reason to suggest that the supply of sediment to this part of the Bay will cease and in consequence sand will continue to accumulate there. It seems unlikely that future recurves will grow out across Sutton Creek towards the Howth mainland, for current scour in the deep creek is too great. It is more probable that they will grow along the sand bar bordering Sutton Creek, as shown in *Figure 2:2*.

Sedimentology

Sedimentary characteristics lead to deductions about past and present conditions of sedimentation.

Grain size, the degree of sorting into size classes, and the sphericity and roundness of particles are those features which offer most information.

Sediment can be described and classified according to its mean particle size. Accepted terminology passes from boulders through cobbles, pebbles, granules, and finally to sand, silt, and clay. The sand class can be sub-divided into five grades – very coarse, coarse, medium, fine, and very fine.

It has already been stated that, under the influence of Flandrian currents, sediment flooring the Irish Sea has been sorted so that in the Dublin Bay area sand has accumulated. Throughout the Bay the sand is of the fine grade, but it is a little coarser over some areas of the morainic highs in the Bay approaches, i.e. the Kish and Burford Banks.

A number of samples taken from the various sub-environments on Bull Island show the predominant sediment type there also to be fine sand; that is, the grain diameters lie in the 0.125 – 0.250 mm range. Moreover, diagrammatic triangular plots of the sediment show that although the sub-environments can be distinguished readily by their physiography, their sediments bear close similarity to each other. Only in parts of the *Salicornia* flat can the sediment be termed mud as opposed to sand or mud-sand *(Fig. 2:4)*.

There is little material on Bull Island that is coarser than the fine sand grade. Most of the coarser material that does occur has been derived either from the comminution of shells or from the raised beach and glacial gravels on the Kilbarrack shore.

Plate 2:4. Rhomboid ripples (bottom right) at the base of the swash zone grading into straight-crested ripples in the proximity of a runnel. Sea is to the top, left.

(C. R. Harris)

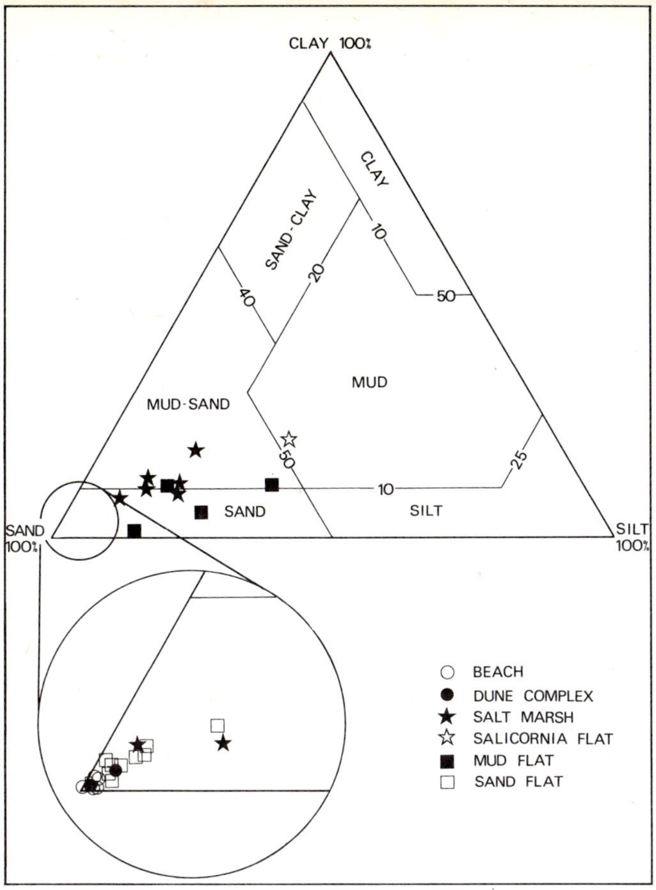

Fig. 2:4. Triangular-plot diagram of sediments from North Bull Island. Clearly demonstrated is the homogeniety of sediments from all the sub-environments.

Another important parameter in sedimentology is the degree of sorting that the sediment exhibits. A sample is well sorted if all its constituent grains are approximately the same size. A poorly sorted sample contains grains having a great diversity of sizes. Material from the beach, dunes, and lagoonal sand flat is all very well sorted; but on the salt marsh and the lagoonal mud flat, where silt and clay account for more of the sediment, sorting is poor. On the *Salicornia* flat, which has the greatest percentage of fine sediment, sorting is very poor.

The final sedimentological parameter to be discussed is that of sphericity and roundness of quartz grains. Quartz, a silica mineral, is the dominant constituent of sand and is conventionally used for textural studies because of its stability and abundance. Roundness deals with the angularity of a grain's corners while sphericity deals with the degree of approximation of a grain to a sphere. Between them, they provide information about the transportation and textural maturity of the sediment. The concept of textural maturity is based on the extent to which sand is free from mud and is well sorted and rounded. Randomly selected grains from all sub-environments on Bull Island show a high degree of roundness and sphericity and, furthermore, that there is little textural variation between the sub-environments. This homogeneity firstly suggests a mature stage in the sedimentary history, caused by long transportation times during which selective sorting has operated at a high level, removing the finer materials. However, the poor sorting of the muddier sediments infers a lesser degree of maturity in which winnowing has not removed the muds. This appears to contradict evidence from the rest of the island. In fact the muds are from a different (local) source than the sands. The latter are unimodal in derivation, having originated as Pleistocene drift in the Irish Sea, which underwent transportation and sorting. Homogeneity also suggests that the sub-environments are too young to have imprinted their own characteristic shape potentials on grains. For instance, in long-established dune environments, the grains become more rounded and spherical than beach sands.

The mineralogical composition of the Bull Island sediments can be dealt with briefly. Of the sand-grade material, the predominant mineral is quartz (90%), while most of the remainder is calcite, derived from local shells. Sand can be described according to the dominant mineral or minerals. In the case of Bull Island, therefore, it is a quartz- or silica-sand.

Sedimentary structures

Sedimentary structures are surficial or internal. Of the former, the most common are ripple marks *(Plates 2:4, 2:5, 2:6)*. The beach on Bull Island provides an excellent arena for the study of a range of wind and water formed ripple marks. The backshore, necessarily, is operated on by aeolian factors, individual sand grains moving in a series of hops down the air stream. Commonly wind ripples are seen, recognised as such by the accumulation of less dense shell hash and organic debris along their crests. These ripples are generally asymmetrical, their steep side facing down wind and their crests lunate. With increased wind speed, two other bedforms can develop. The more spectacular of these are small-scale barchan dunes, a few centimetres in height and about a metre between the wing-tips. Groups of these have been seen moving 'en echelon' along the beach at a rate of several centimetres per hour. Entrainment of blown sand also occurs as windrows – stringers of sand moving in the wind direction.

Plate 2:5. Straight-crested ripples on ridge becoming cuspate in a channel (bottom, right) due to increased depth and flow.

On the foreshore, the area between the berm and the first ridge demonstrates the relationship of ripple types with beach slope and landward decay of waves. Immediately below the high-water line, where the beach slope is five to six degrees, occurs the swash-zone. Here, the ripples are very flat. Their recognition lies in the accumulation of organic debris and shell hash in their troughs (note that this feature of water-formed ripples is in contra-distinction to wind-formed ones) which form sinuous colour-bands about 50 cm apart. These are swash-zone ripples and occur in a band a few meters wide along the breach.

Seaward of these, where the beach slope decreases, flat, rhomboid ripples occur, formed by the backwash. As the proximity of the runnel is reached and where water depth increases, straight-crested, small-scale ripples are found *(Plate 2:4)*. These are aligned parallel to the beach and have a wavelength of about 8 cm and a height of about 1 cm. Finally, in the runnel where water is deepest, stream flow creates a more chaotic arrangement of typical linguoid ripples which are concave downstream *(Plate 2:5)*.

On the tops of the ridges, straight-crested, small-scale ripples occur. At low tide, a sheet of standing water often covers this surface. Under suitable conditions, such as a strong south-westerly or north-easterly wind, the surface of the water is agitated sufficiently to form interference ripples in the sand cresting normal to the main set. This produces a ladder-backed ripple pattern *(Plate 2:6)*.

Due to the homogeneity of the material that has been deposited as beach and lagoonal sediment, internal structures such as laminations, ripple-marks, and cross-bedding are not clearly apparent unless special techniques are used. For example, from pits dug in the foreshore the only structures immediately visible are shell bands. An exception to this lies in the dune environment. Here, wind etching of eroded dunes has emphasised the typically chaotic and cross-bedded nature of dune deposits.

In the salt marsh, very distinct internal laminations are visible, representing alternations in sand and mud deposition. At the entrance to the lagoon, and near the tip of the 1936 recurve extension a core shows interesting mud/sand laminations *(Plate 2:3 c)*.

At the north-eastern end of the island, patches of very soft sand exist. Their position is shown in *Figure 2 : 3*. Below the 2 cm thick crust of smooth or rippled sand the sediment here assumes a vesicular 'sponge-cake' texture to a depth of 30 – 40 cm. The diameter of the largest vesicles reaches 4 mm and near the surface they are spherical. However, at greater depth the vesicles are smaller (1 mm diameter), often composite, and generally flattened in a horizontal plane due to loading by the sediment above. This vesicular structure originates in the following manner. As the intertidal sand is drained during the ebb-tide the intergranular spaces become filled with air. With the flood-tide, water inundates the surfaces faster than the more sluggish interstitial water can saturate the sediment and replace the intergranular air. The air becomes trapped in the sediment between two bodies of water and vesiculates, because a spherical shape encloses the greatest volume whilst presenting the smallest surface area. The bubble-shape is retained during sub-aerial exposure of the beach through surface tension of the surrounding grains. Constant tidal inundation tends to enlarge the vesicles nearer the surface whilst sediment loading causes collapse and coalescence of deeper ones. The eventual disappearance of vesicles at depth is caused by the inhibitory effect of saturated sand, rather than by complete flattening.

While vesicular sand patches are a common feature of beaches, certain conditions are requisite. For example, the sediment must be of sand grade, the slope of the surface under which they form must be less than eight degrees, surf activity must be minimal, and the area must be in the higher intertidal zone to permit maximum sub-aerial exposure and drying time.

Conclusion

North Bull Island appears to be at a most interesting stage in its development. Probably the most spectacular aspect for the coastal geomorphologist is the pattern of recurves. At present they are at a most critical point in their evolution as they provide a classic example of the growth of one recurve with the concomitant destruction of another.

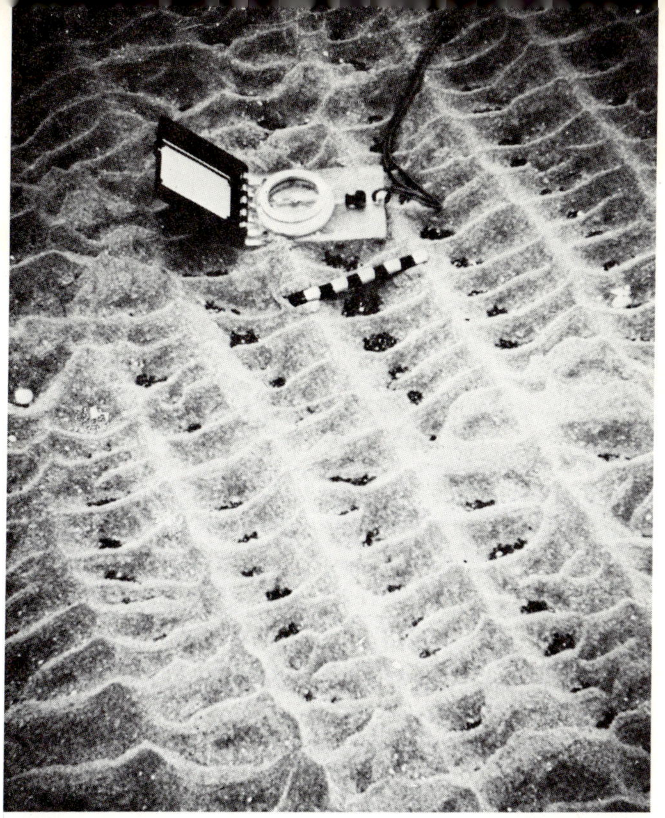

Plate 2:6. Ladder-back interference ripples. Scale graduated in cms.

In so doing, they allow close study of the natural processes acting on spit growth and recurve formation, which are still only partly understood.

Perhaps the most obvious geomorphological feature is the presence of dune ridges which can be related to old shorelines and dated, thus providing a sensitive tool for the study of the island's growth.

The island also provides examples of the response of natural environments to new factors operating on them. This is exemplified by the spread of the mud and *Salicornia* flats over the lagoon, as approach towards equilibrium with new conditions made operative in the lagoon by the advent of the causeway. The muds are accumulating at a rate of between 2 and 5 mm per year and similar figures have been computed for the adjacent sand flat, resulting from studies of temporal changes in ostracod and foraminiferal assemblages. A longer term response to natural conditions is the steady and inexorable growth of the salt marsh over the lagoon.

Finally, the beach can be used not only as a recreational area, but also for the study of a variety of ripple-marks, each a response to the hydrological conditions acting on that particular point.

Acknowledgement.

Geological investigation of North Bull Island was suggested to the author by the work of Dr. R. T. R. Wingfield, now of the Institute of Geological Sciences, Leeds, who initiated sedimentological studies of the island and adjacent lagoon whilst at Trinity College, Dublin.

Chapter 3

COMPARISON OF THE HABITATS

D. W. Jeffrey *Department of Botany Trinity College*

The various sub-environments whose historical growth and development have been considered are characterised by changes in physical properties and in a variety of chemical and biological processes. Some of these characteristics set out in *Table 3:1* are obvious but most must be detected or studied under laboratory conditions. The list is by no means exhaustive, but serves to illustrate the complexity of the environments. This complexity highlights the difficulty of distinguishing major from minor ecological factors, and indeed of setting aside factors of no relevance to the distribution or performance of living organisms.

The intertidal zones

The tides

All who bathe or sail, dig for lugworm bait, or wish to study the ecology of estuaries, must take account of the tides The theory governing the complex movements of tidal waters need not concern us, but an overview of tidal rhythms is helpful. Three types of regular rhythm may be distinguished using a tide-table *(Fig. 3:1)*. The most obvious is the roughly twice a day cycle of high and low water. Parts of the intertidal flats and the beach are thus flooded with sea water twice in twenty-four hours, and then drained, exposing a surface to air and possibly to non-saline water in the form of rain or drainage water. It is interesting to note that the curves of daily tidal cover are asymmetrical in the Bay. The time for the tide to flow from low to high water is approximately 6 hours 40 minutes, whilst the ebb takes but 5 hours 40 minutes. This may ensure that the northern lagoon will always remain open at its mouth.

The second type of rhythm is an oscillation in tidal range which occurs twice a lunar month. In this cycle, a period of high tidal range occurs at full and new moon, the difference between high and low tide then being about 3.6 m in Dublin Bay. This is followed by a period of low tidal range of approximately two metres *(Fig. 3:2)*. The high range tides, known as spring tides, usually lap up to the four metre O.D. mark, and retreat to about 0.4 m O.D. Ordinary spring tides reach the upper parts of beaches and cover most of the salt marsh. In retreating, large areas of flat are exposed, and permit investigation of the lowest intertidal areas.

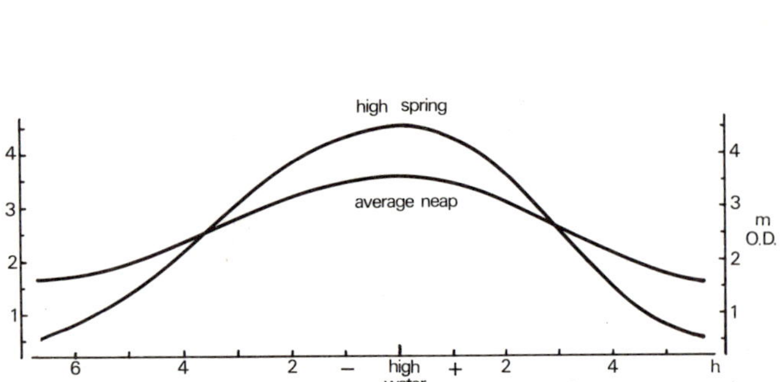

Fig. 3:1. Diagram illustrating the patterns of tidal cover in the course of a single spring or neap tide in Dublin Bay. Note that the ebb part of the cycle is about one hour shorter than the flow.

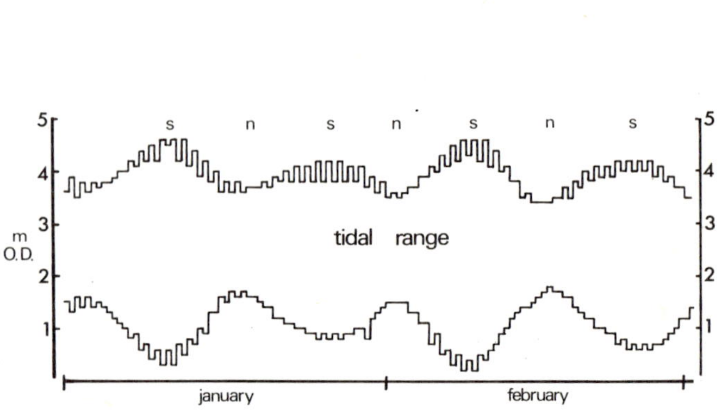

Fig. 3:2. Heights of all high and low tides for Dublin January and February 1974 as predicted by tide table. This diagram indicates the rhythmic alternation of Spring (S) and neap (N) tidal cycles, and provides a method for computing tidal cover or exposure of a given portion of the island.

Table 3:1 Environmental factors in four major habitats.

Environmental Factor	Beach	Dune complex	Salt marsh	Lagoonal sand and mud flats
Sediment type	Gravel to fine sand	Fine sand – wind blown	Fine sand to clay	Fine sand and silt
Source of organic matter	Tide line debris	Higher plant growth	Tide line debris Sedimented fine detrital particles Algal mats	
Tidal immersions per year	About 5 to more than 700	Tidal incursions rare and catastrophic	Approximately 5 to 250 Covered by spring tides only	Less than 250 to more than 700
Interstitial salinity	Variable only with rainfall	Low – result of wind blown spray	Complex interaction between fresh water table, tidal cover, and rainfall	Tidal water may be diluted by drainage water and rainfall
Carbonate content	Present in all sediments from comminuted shell fragments Loss by leaching with age			
Aeration	Aerobic conditions at least at surface	Transient reducing state in flooded slacks	Reducing condition common, associated with fine sediment, organic matter, and summer temperatures	
Sulphur metabolism	Generally as sulphate		Sulphides and sulphates Sulphur possible	
Phosphate status	Occurs as precipitated apatite minerals throughout Loss by leaching with age			Accumulation by algal mats
Nitrogen status	Low – debris and detritus are main sources	Active biological fixation by symbiotic and free-living organisms	Detritus and debris	
Heavy metals	Generally present in trace quantities		Accumulated by biological processes, precipitation, and absorption by fine sediment	

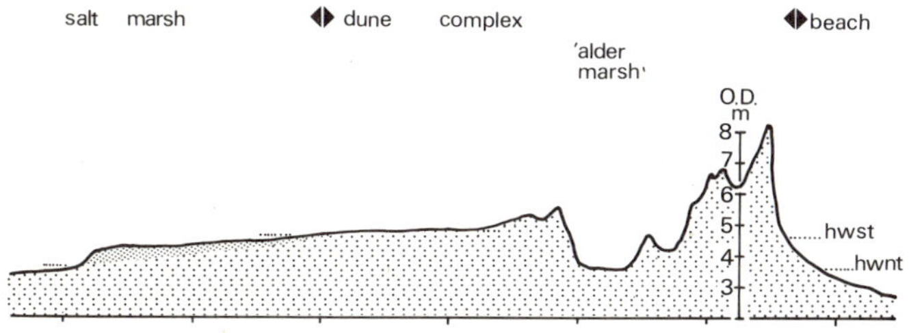

Fig. 3:3. Topographic section of the island through the Alder marsh. The vertical scale is exaggerated 25 times. The wedge of salt marsh sediment is seen to be overlying a sandy core. Four dune ridges are evident, with the floor of the alder marsh close to the high water neap tide mark. This is also the point where the salt marsh sequence begins.

The low range neap tides cover beaches and salt marshes only to about 3.5 m O.D., and expose flats down to about 1.5 m O.D. These tidal cover ranges may be compared with the topographic profile in *Figure* 3:3. It is to be noted also how the time of exposure or immersion of a given point will depend largely on the state of this twice monthly cycle.

The third regular rhythm is one of spring tide ranges that vary through the seasons of the year. The highest spring tides are thus expected at the equinoxes, in March and September. These highest of all tides are important in that they impose a certain degree of salinity at least to the 4.7 m mark on beaches and salt marshes, and are thus an important point of ecological demarcation. This point is frequently indicated by a quite massive drift line of organic debris.

The rhythms described, attributable to astronomical forces, are modified very much by wind action. In this fashion the astronomical high tide of January 1974 was backed by a south-easterly gale, and waves broke against the main dune ridge, eroding it and leaving cliffs of sand, up to 4 metres high. Dune slacks were not flooded on this occasion as happened in 1925 when only one rudimentary ridge stood between the Alder Marsh and the sea.

The ecological effects of tidal inundation are appreciated when sea water is compared with terrestrial river water or rain water *(Fig. 3:4)*. Sea water is roughly a thousand times more concentrated than river water, and there are also differences in the proportions of ions in the two solutions. Sodium, magnesium, chloride, and sulphate are the four most common ions in sea water, whilst in rivers calcium, sodium, bicarbonate, and sulphate head the list. Bromide ion, which is very rare in river waters, is present at concentrations up to 100 p.p.m. in the sea, whilst silicate, a ubiquitous terrestrial ion common in rivers, is at low concentrations in the sea. These contrasts indicate the size of physiological change that can occur to plants and animals with the passing of the tidal cycles. They also indicate the magnitude of environmental difference between the central dune complex, with its fresh water table and the flanking beach and salt marsh.

Aeration, microbiology, and geochemistry

The deposition of fine sediment in still water has profound effects on all living organisms of intertidal environments. The admixture of silt particles with fine sand leads to a medium with finer pores and a diminished capacity for water to drain from them. Diffusion of oxygen is ten thousand times slower in water than in air, and hence if oxygen is consumed by aerobic organisms, it may not be replaced. The presence of organic matter leads to an intensification of biological activities including oxygen consumption. The development of oxygen free (or anaerobic) conditions has consequences in terms of chemistry, microbiology, and in the mode of life and activity of plants and animals. One way of thinking about these rather complex matters is to imagine a column of muddy sand with more anaerobic conditions

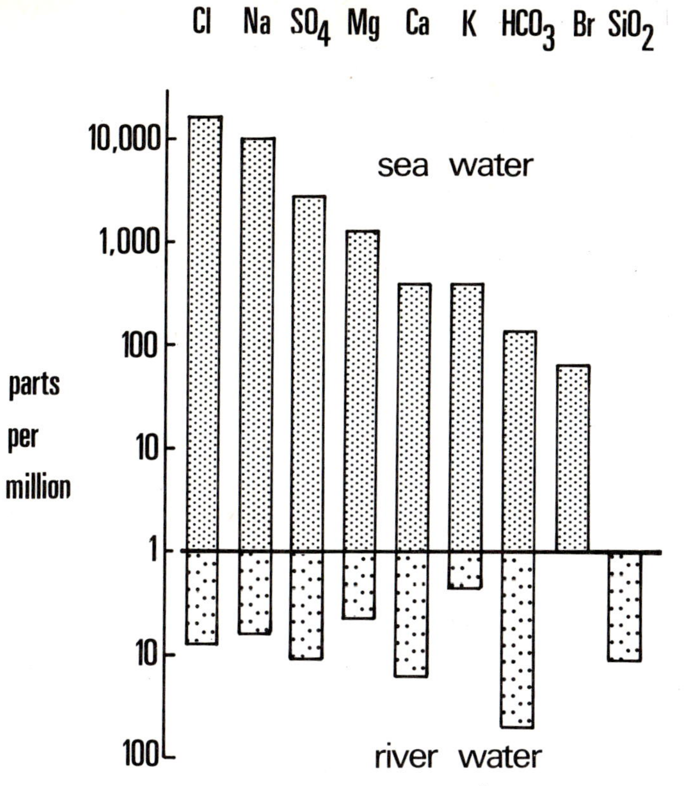

Fig. 3:4. Comparison of major ionic concentrations in typical sea and river waters. The scale is logarithmic. Sea water is roughly a thousand fold more concentrated, with only silicate ion more plentiful in river water.

developing with depth. The upper layer would typically be brown turning to grey brown and then black. The thickness of the upper brown layer is a good index of both the porosity of the sediment, and its organic content. High porosities and low organic matter contents produce the deep golden sands of a good beach. A blackened sand or mud surface bespeaks either a very silty sediment or one with much organic matter. The schlick muds close to the causeway have both.

Mud colour is a direct consequence of the iron minerals in the mud, brown mud containing hydrated iron oxides, with increasing proportions of iron sulphides giving the colour changes through grey-brown to black. The conversion to sulphides is dependent entirely on the activities of bacteria such as *Desulphovibrio* sp. and *Desulphotomaculum* sp. These organisms link the utilisation of energy in organic matter not with gaseous oxygen consumption, but with the conversion of sulphate to sulphide. It must be remembered that sulphate is readily supplied by sea water *(Fig. 3:4)*. The interstitial water thus becomes sulphide saturated and metal sulphides of low solubility are precipitated. Iron is the dominant element to be so precipitated, but zinc, copper, and lead accumulation in anaerobic sediments also occurs. The schlick muds contain twice as much zinc and copper as better aerated sediments *(Fig. 3:5)*. Evolution of gaseous hydrogen sulphide (the smell of bad eggs associated with low tide at Butt Bridge on the Liffey) together

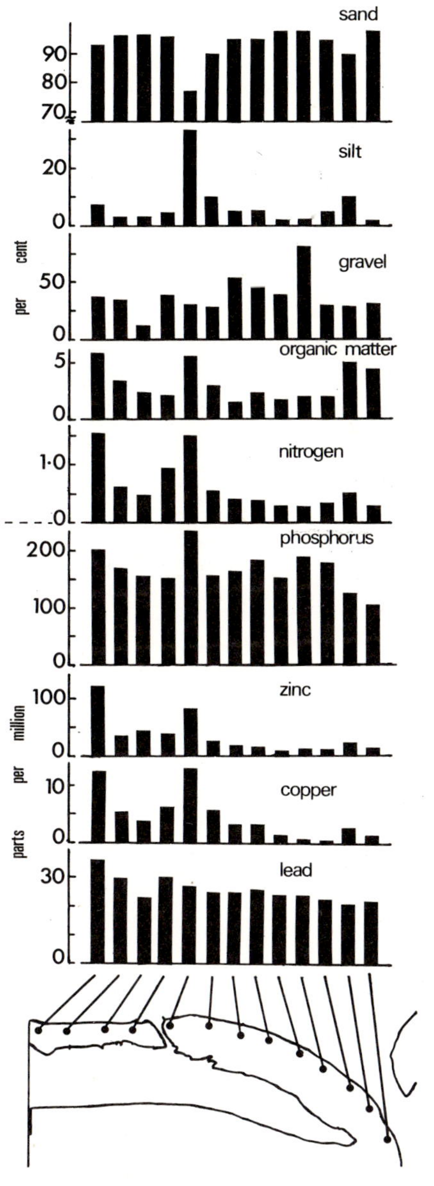

Fig. 3:5. Physical and chemical analysis of inter-tidal sediments. Each bar represents the mean of five samples (0-10 cm) collected 20-100 m from the sea wall.
Sand and *silt* are complementary, sand on the whole predominating.
Silt is of importance on the N.E. side of the causeway for approximately 2 Km, and probably originates from a fluvial source.
Gravel outcrops occur off Kilbarrack but is elsewhere overlain by the finer sediments.
Organic matter which contains most of the *nitrogen* in a sediment is strongly related to the deposition of silt. Except for the site to the N.E. of the causeway *phosphorus* figures are not exceptionally high, there being a tendency for much higher concentrations to occur to the S.W. of the Bull Wall. In absolute terms no heavy metal reaches anomalously high concentrations. However, it is clear that these metals are allied to the deposition patterns for fine sediment and organic matter. *Zinc* concentration should be monitored, and the presence of additional metals verified.

with ammonia, methane, and carbon dioxide are all characteristic of the anaerobic situation. Especially intense reducing activity high on the shoreline may be associated with a white surface film. This film indicates the presence of another bacterium, *Beggiatoa alba,* which completes the reduction of sulphide to elemental sulphur. This may often be seen close to the outfalls of the Santry River or Nanikin Brook.

These same reduction processes are still observable on salt marsh, but greatly localised by the pore systems created by plant roots, and by the aggregation of fine sediment particles high on the marsh. Aeration here tends to be lowest in the narrow *Puccinellia* zone.

In contrast to the abundance of sulphate referred to above, phosphate, a key ion in all living organisms, is normally present in sea water at very low concentrations. Hence the primary accumulators of minerals in the mud flat ecosytem, the algae, are effective at absorption and accumulation of phosphate from tide water. Algae are buried by continued sedimentation and then decay. Phosphate released by this decay, usually under anaerobic conditions, is precipitated on the surface of sedimentary carbonates, to form a family of calcium phosphate minerals, the apatites. Their formula approximates to $Ca_{10}\,(PO_4 . CO_3)_6\, F_{2-3}$ and their presence serves both as the major phosphate source to rooted salt marsh plants and as a buffer to the concentrations of phosphate in open water. Absorption of phosphate by algae and transfer to sediment is a major process which is exaggerated in the eutrophication of estuaries and which we need to understand in greater detail.

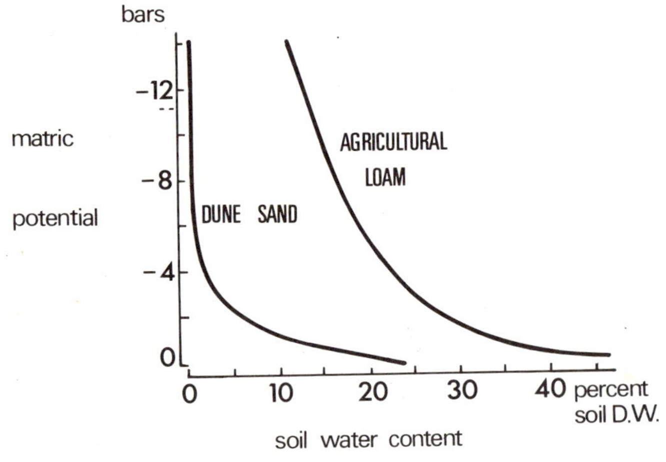

Fig. 3:6. Soil moisture characteristic curves for dune sand and an agricultural loam. A loam will supply 60% more water to a plant before wilting than the sand.

The dune complex

Whilst the intertidal environments of beach, sand flats, mud flats, and salt marsh may be thought of as overlapping and intergrading, the sand-dune complex stands somewhat apart. We have seen that the growth of the island in length and breadth has resulted from accumulation of a remarkably uniform material. Once above the reach of the highest tides however, this mass of fine sand represents a very different environment for plants, animals, and micro-organisms from any so far considered. Its instability is of a kind different from that of the intertidal areas. For example, wetting and drying of the surface by rain and dew versus evaporation by sun and wind is an almost cyclic counterpart to the rhythm of tidal inundations in the other environments. The dune system is virtually isolated from the sea as a source of organic matter, minerals, and organisms. Much blown sand is intercepted by the outer ridges, and the only gale blown salt spray imparts a light transient salinity to the leeward dunes. Rainfall on the dunes maintains the lens-like fresh water table.

Here exists a self-contained set of habitats, in which differences are clearly related to age. In this respect there is a resemblance to the zonal differentiation of salt marsh, and in both dune and salt marsh systems a time sequence of soil development is demonstrable. This is especially clear if the upper 10 cm of the sand from each dune ridge is sampled. On the first ridge, barely 20 years old, sand may be easily scooped up in the open hand, and will run between the fingers. It is pale yellow in colour with a few flecks of organic matter from marram grass litter or even seaweed. On the next little ridge, bare sand is less easy to find, is a much darker grey, and sand grains tend to be aggregated. On the far side of the Alder Marsh, on the oldest part of the island, fingers are not good implements for digging holes because of a mat-like layer of humus with virtually no bare sand. In a little under two hundred years quite a reasonable soil has developed from the same material that forms the beach and indeed covers the sea floor off the Wicklow coast.

Nitrogen fixation

Laboratory analysis of this series of soils confirms the visual and textural impression that organic matter is accumulating with age, and with it substantial quantities of nitrogen. Experiments have shown that nitrogen supply is a key limitation to the growth of dune vegetation, and the process of biological nitrogen fixation is thus of considerable interest. At least four different systems for nitrogen fixation are present in some part of the dune complex, and it is feasible that others exist.

Most conspicuous, and certainly of prime importance are the leguminous plants. Their nodules are active through the summer months, decaying in autumn to release organic nitrogen to the soil. It is certain that other plants effectively tap this supply. Yellow rattle *(Rhinanthus minor)* does so directly as a hemiparasite with haustoria clamped to legume roots. The grasses and orchids absorb nitrogen and other nutrients via a mycorrhizal root system that may be able to assimilate organic nitrogen (*Plate XVIII*). The rate of legume-nodule fixation is probably closely related to water supply, and is thus especially active in those slacks which are closest to a water table without being flooded.

Two other fixation systems are visible with the naked eye. A common thallose lichen in 30–40 year old dunes is a species of *Peltigera* having a blue-green algal symbiont. Here too fixation activity is probably limited by the water content of the thallus, and may well continue throughout the winter months. The woody nodules of the Alder show fixation activity at least into autumn, but which probably ceases on leaf fall.

Free living blue-green algae of the nitrogen fixing *Nostoc* group are classic dune inhabitants. Occasionally colonial forms, like small green gelatinous spheres or buttons, are obvious in damp weather. Otherwise their detection in the thin crusts that form on top of sand surfaces is a laboratory task.

An enigma of the dunes is the source of nitrogen for *Ammophila,* the plant which literally holds together the main ridge without either plant-litter, or obvious co-existing nitrogen fixers such as the lichens or legumes. The answer may be a free living bacterium living in close proximity to the roots, or even a bacterium associated with leaf surfaces.

Water in the dune system

Lack of water has already been given as a reason for the limitation of nitrogen fixation, and the lens-like fresh water table lying below the dunes has been described. Why are the dunes a recognisably dry environment when they receive essentially the same rainfall as the neighbouring mainland? The principal reason is the size of the sand grains, and the related interstitial spaces. Water remains in a sand or any other soil, mainly because spaces between particles act as capillary tubes. The smaller the particles, the greater the retaining force because the capillaries are also small. Most soils have a whole range of particles from clay (2 nm diameter) to gravel (greater than 2 mm diameter) and a whole range of capillary sizes in what is known as the soil matrix. In dune soils, the matrix is very simple, with only the relatively large sand grains plus varying amounts of organic matter to be taken into account. As any soil dries out, water is removed firstly from large capillaries, and then successively from finer ones. The force retaining water rises all the time as the percentage of water in the soil drops. This relationship may be expressed as a curve which characterises a particular soil. In *Figure 3:6* Bull Island dune sand is compared with an agricultural soil, with the retaining force (matric potential) expressed in bars. It will be seen that water leaves a dune soil very easily, and rapidly reaches a matric potential at which plants wilt, i.e. some – 20 bars. Dune soils possess neither a very high capacity for water storage, nor the ability to retain water at useful potentials for utilisation by roots. The permeability of sand is also high, and free standing water is only seen in winter when occasionally the water table rises above the level of the Alder Marsh.

Chapter 4
DISTRIBUTION AND BIOLOGY OF ALGAE

P. H. Pitkin *Botany Department Trinity College*

Plants of intertidal flats

On the intertidal flats the dominant plants are green algae and glasswort, *Salicornia* spp (*Plate X*). The green algae are almost ubiquitous on the flats between Fairview and Sutton Creek and in places form a mat which can be seen at low tide from the coast road. *Salicornia*, a flowering plant, grows in dense communities on areas of firm sediment north of the causeway extending for about 2 km towards Sutton (see map *Fig. 11:1c*). Beside the green algae and the *Salicornia* clumps of *Fucus* are conspicuous from the coast road at Kilbarrack attached to stones in the mud. Much less conspicuous but botanically interesting, are the marine flowering plants *Ruppia maritima, Zostera zana* and *Z. noltii*, which occur sporadically north of the causeway, the *Zostera spp.* mainly at Kilbarrack. This chapter deals particularly with the green algae (Chlorophyta) because since 1973 they have been closely studied and because they are undoubtedly an important part of the ecosystem. They are eaten by herbivorous animals, in particularly large quantities by the wildfowl which visit the North Bull in winter. On decomposing they break up and provide food for detritus feeders. Their matted and tangled growth forms trap suspended mineral particles and probably assist in the deposition of sediment.

The algae

Many of the green algae found on the Dublin Bay mud-flats belong to the order Ulvales, most of the members of which have a thallus consisting of a sheet or tube of cells. Species of *Enteromorpha,* which are normally tubular, are particularly common. This genus is often thought to be characteristic of brackish water and especially the polluted water of many harbours and estuaries where *E. compressa* and *E. intestinalis* are usually abundant. However, there are other species of *Enteromorpha (E. prolifera, E. ramulosa)* which are characteristic of unpolluted mud-flats and are an important part of the vegetation of the lower regions of salt-marshes. These plants are long, narrow and finely branched. They form tangled masses and could easily be mistaken for filamentous algae. *Ulva lactuca,* the common sea lettuce, is a member of the Ulvales in which the thallus is a sheet two cells thick. On rocky shores this is a delicate plant, having a frond with a definite shape and a wavy margin, growing attached to the bottom of rock pools. On the mud-flats the plant is unattached, firmer in texture and very ragged. The commonest filamentous algae on the mud-flats are species of *Rhizoclonium, Cladophora* and *Chaetomorpha,* members of the Cladophorales. Their filaments are made up of multinucleate cells which appear to be lined by net-like chloroplasts with numerous pyrenoids.

The fact that many of the green algae are difficult to identify is an obstacle to the study of their ecology. There is no up-to-date taxonomic work which covers the whole of the marine Chlorophyta and there have been many descriptions of species of *Rhizoclonium, Ulva* and *Cladophora* published which are no longer recognised. The numerous species of *Enteromorpha* are particularly difficult to identify. Since their growth forms vary considerably, the only useful characters for separating them are microscopic ones: the dimensions and arrangement of the cells, the number of pyrenoids and shape of the chloroplasts and in some cases the microscopic characters of the branchlets. Specimens must be examined immediately after collection, mounted in diluted sea water and stained with a weak solution of iodine. Bliding's study of the genus (1963) is the only satisfactory means of identification, although it lacks a key and contains almost no information on the habitats of the species. Bliding's

I. Boundary between lagoonal flats and salt marsh. The small cliff marking the high water neap tide mark, here is stabilised by green algae. At the edge of the creek *Halimione* may be seen, and the striking Autumn colour of *Salicornia* is evident amongst *Puccinellia*.

(C. H. Holland)

II. Beach near the Bull Wall showing the accumulation of debris, mainly green algae, but also tubes of the sand mason, *Lanice conchilega*. The gulls are feeding on animals such as *Hydrobia ulva* amongst the detritus.

(B. Healy)

I C. H. Holland II B. Healy

III C. Harris IV D. W. Jeffrey

species and subspecies are based on the compatability of gametes and for this reason are objective 'biological species'. In my own studies of *Enteromorpha* I have followed Bliding's treatment.

The flats on which algal communities are the main type of vegetation extend vertically from the approximate level of low water neap tides to the level of the miniature cliff at the lower edge of the salt-marsh proper, at about 3.7 – 3.9 m. above ordnance datum (*Plate I*). The tide, measured at the North Wall, Dublin, fails to reach 3.7 m above O.D. on only about 30 days each year so most of the algal habitats are covered by the tide twice every day. As the tide comes in, much of what at low tide is often described as the 'algal mat' floats free from the substrate. Even when the sea is calm the high tide redistributes the algae. This is easily demonstrated by clearing marked areas. In December I have found cleared areas 2 m. square to be unrecognisable after only a day. Similarly, stakes driven into the mud will retain hanks of drifting alga as the tide recedes. In the winter after the November and December spring tides much algal material from the mud flats can be found on the salt-marsh in pools or clinging to plants.

Most of the *Enteromorpha* in its young stage seems to be attached. It is quite common to find young plants of *E. flexuosa* in early spring, growing on pieces of debris, on *Hydrobia* or cockle shells and on sand grains, but these objects are too small to anchor the plants for long. Sporelings of *Enteromorpha* are also found attached to the tangled wefts of older algae. In the life history of most species of *Enteromorpha* and *Ulva* there is a regular alternation of morphologically similar haploid and diploid generations. The haploid generation, by mitosis, produces gametes with two flagella; the diploid generation, by meiosis, produces zoospores with four flagella. Normally gametes will only fuse with those of a plant of a different strain, but there is a visible difference between 'male' and 'female' gametes. As far as is known, most of the Cladophorales have a similar type of life history, though in some, the haploid and diploid generations are not identical. *Rhizoclonium implexum* and *Blidingia marginata* both of which grow among *Salicornia* seem to retain close contact with the substrate throughout their life.

III. The form of the new spit as seen in 1972 against a rising spring tide. The rudimentary colonisation of the spit by hummocks of vegetation *(Agropyron junceiforme)* is striking. *(C. R. Harris)*

IV. Leeside of the first dune ridge showing the trapping of newly blown sand by marram grass, *Ammophila arenaria*. *(D. W. Jeffrey)*

Distribution of the algae

Despite the movements of the algae with the tide, there seems to be a consistant pattern to the distribution of the algal mat at low tide between Fairview and Sutton. *Fig. 4:1A* shows the estimated dry weight of alga/m^2 within 100 metres of the shore, at twenty stations where samples were taken in July and August 1973. At each station alga was collected from 10 cm squares along a line, 20, 40, 60, 80 and 100 m. from the sea wall. The length of this line which crossed patches of algae was also measured *(Fig. 4:1B)*. Between the mouth of the Tolka and Bull Wall the cover and biomass are small. Immediately to the north of the Bull Bridge the amount of alga increases, and along the shore between the bridge and the causeway the cover and biomass of the algal mat are at their greatest. North of the causeway patches of alga are less extensive and less frequent. These measurements correspond approximately to the observed distribution of algae on the mud flats further away from the shore. The whole of the area between the bridge and the causeway are covered by large areas of lush green, while on the other side of the causeway the algal growth is much less noticeable.

In *Fig. 4:1C* the occurance of sone individual species of algae is shown for each of the twenty stations as recorded on this survey. *Enteromorpha compressa* and *E. intestinalis* extend the length of the shore. Near the mouth of the Tolka the former is the dominant alga and is apparently tolerating a high level of pollution. *E. prolifera* also extends the length of the shore but, like *Ulva lactuca,* was found in greatest quantity and is dominant, between the bridge and the causeway. Beyond the causeway it decreases in relation to other species. *Enteromorpha ramulosa* was only found in quantity north of the causeway where also *E. ralfsii,* which was only recorded once elsewhere, occurs at four successive stations. *E. flexuosa* occurs exclusively at stations north of the causeway, as does *Percursaria percursa, Sphacelaria pumila* (Phaeophyta), *Polysiphonia macrocarpa* (Rhodophyta), *Rhizoclonium implexum, Ceramium tenuissimum* (Rhodophyta), *Dictyosiphon foeniculaceus* (Phaeophyta) and *Ectocarpus silicuosus* (Phaeophyta) which were each only recorded a few times.

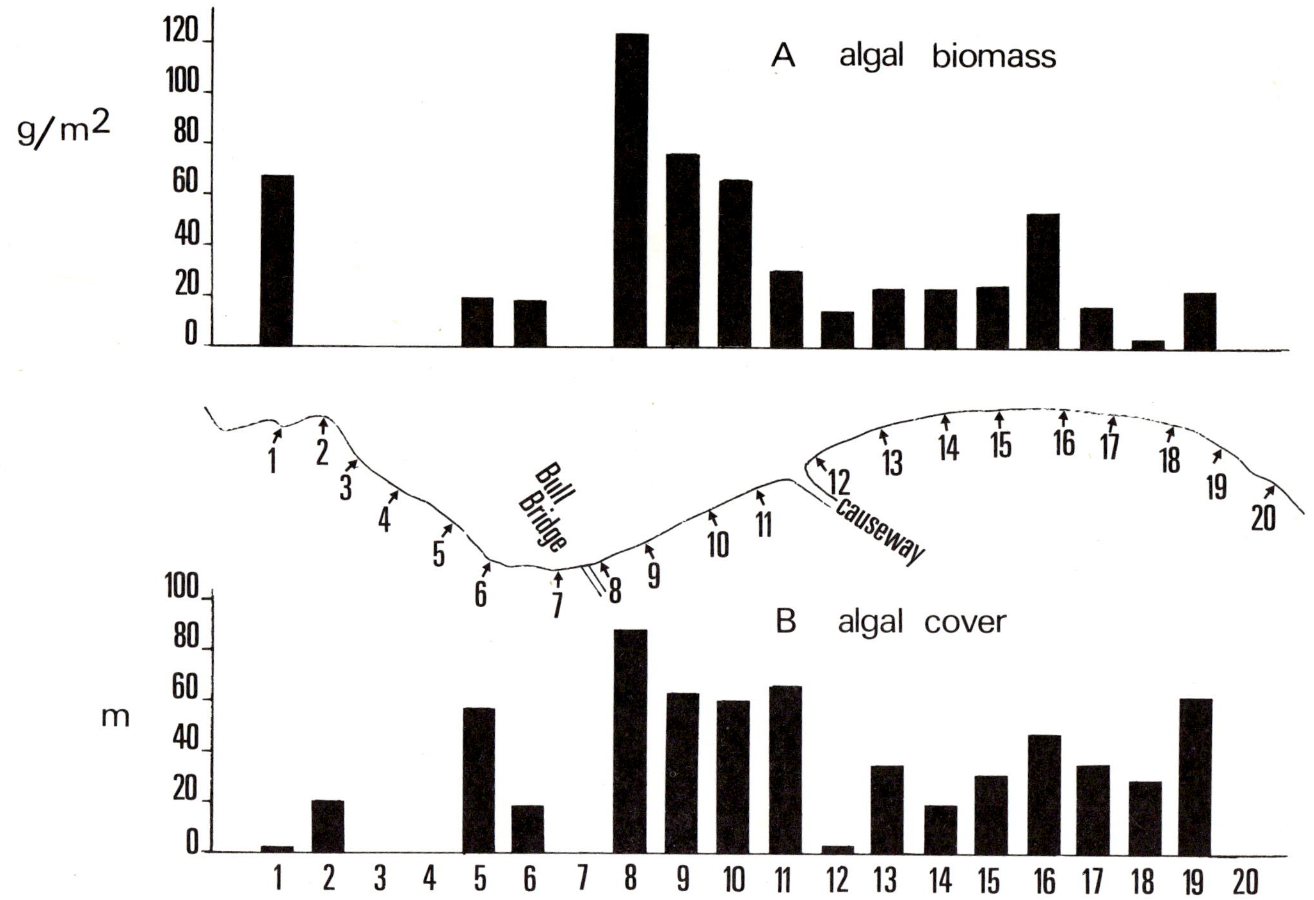

C species distribution

	1	2	3	4	5	6	7	8	9	10	11	12	13	14	15	16	17	18	19	20
Enteromorpha intestinalis	1	5	5	1	3	2	1	5	3	3		2	1	2	1	2		2		2
E. compressa	2					1	3	2			1	1		2	2	3		1	1	
E. prolifera	1	2			1	2	2	4	5	5	5	1	4	2	2	5	3	2	3	1
E. ramulosa		1			1	1	1					3		3	2	3	4	2	2	1
E. flexuosa													1	2	3	4	3	1	2	1
E. ralfsii											2				1	4	2	1		
Uiva lactuca						1	2	2	4	5	1	2	1	1		2	1	3		1
Chaetomorpha linum					1					1	1	1		4	1		1			1

Fig. 4:1. Distribution of algae along the upper 100 m. of the shore at 20 stations between Fairview and Sutton in July/August 1973.
A: dry weight of alga per m², mean of 5 samples from 10 cm², taken along a line at right angles to the shore at each station.
B: length of 100 m. transect, at right angles to the shore, covered by algal mat at low tide.
C: occurrence of the most conspicuous green algae at 5 points along each transect.

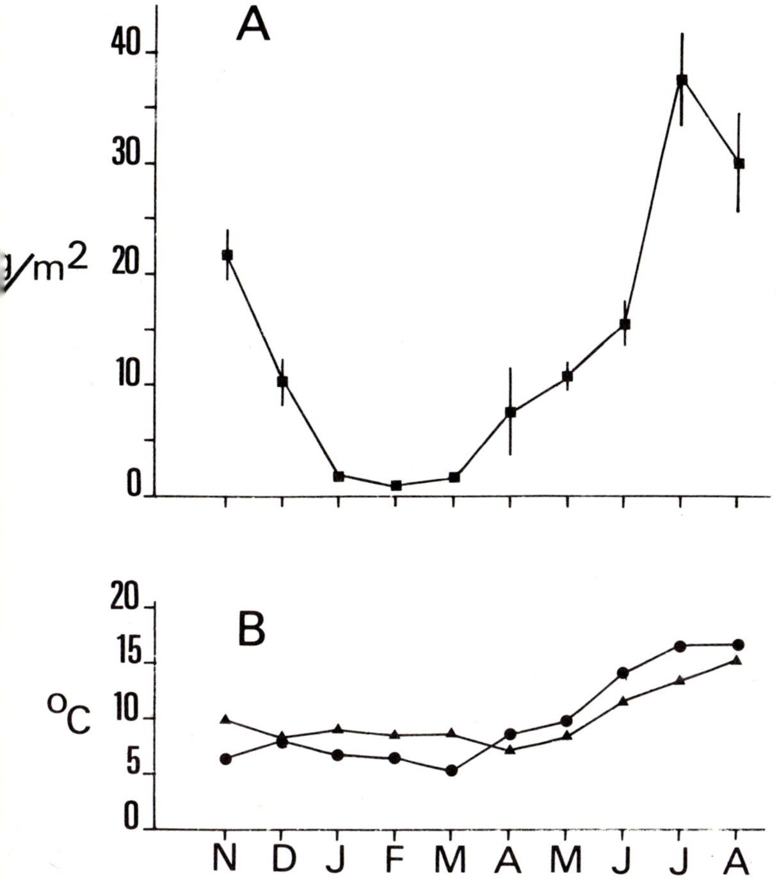

Fig. 4:2.
A: dry weight of alga per m² per month, means and standard error for 52 samples. November 1974 – August 1975.
B: mean monthly sea temperature recorded by the Arklow Light Vessel (▲) and mean monthly air temperature at Dublin Airport (-●), both supplied by the Meteorological Office.

Seasonal changes in algal communities

Seasonal changes in the vegetation of the flats can be observed from the coast road as colour changes. The most striking of these are brought about by the growth and senescence of *Salicornia* north of the causeway. During the summer *Salicornia* forms a bright, pale green carpet (*Plate XI*) which in September rapidly turns to rusty yellow. This colour has usually disappeared by the beginning of November. An equally rapid and only slightly less spectacular colour change takes place in April and May with the new growth of green alga. This is particularly noticeable between the bridge and the causeway.

To investigate more closely the seasonal changes in the algal communities, the dry weight of alga per unit area has been determined monthly, from November 1974 to August 1975, at points along a permanent transect of the mud flats. The transect begins 600 m. north of the causeway and, from the sea wall, crosses the mud flats along a line 126°E of Grid North for 650 m. At low tide it crosses the main drainage channel of the lagoon which also carried water from Santry river, two drainage channels from the salt-march and a patch of *Salicornia dolichostachya* approximately 200 m. wide. The vertical range of the transect is from 2.7 m. above O.D. at the bottom of the channel to 3.9 m. at the foot of the salt-march cliff. At 50 m. intervals along the transect four 50 cm quadrats were placed, varying their position slightly from month to month. Within these, the algal material lying on the surface, excluding drifted *Fucus* fronds, was collected, washed and oven-dried. The monthly means for the whole transect are shown in *Fig. 4:2A*. Seasonal fluctuations in the algal biomass are pronounced. Increases either occur during periods of growth, or when algae are being carried into the area by the tide. Decreases are more complex and appear to be caused by a combination of natural senescence, burial by sediment, feeding by grazing animals and material being carried away by the tide. Sediment is continually accumulating in this part of the mud flats and algae very soon die and decompose after being buried by even a thin layer of fine sediment. *Hydrobia ulvae* is found in very large numbers among all the green algae. I have counted up to 14,400 shells per m² of algal mat. Fronds of *Ulva lactuca* on which *Hydrobia* have been feeding are often found perforated with small holes. Green algae are also an important part of the diet of the large flocks of wildfowl which congregate on the mud flats in the winter.

Between November 1974 and January 1975 the amount of alga decreased rapidly. Comparatively little of this decrease can be attributed to inherent death and decomposition as little senescent material was found at this time of year. It is likely that during this period the growth rate of the algae is slowing down and is insufficient to replace the alga buried by sediment. The greatest probable cause of the decrease is grazing by large numbers of wildfowl which build up on the mud flats in the autumn. The

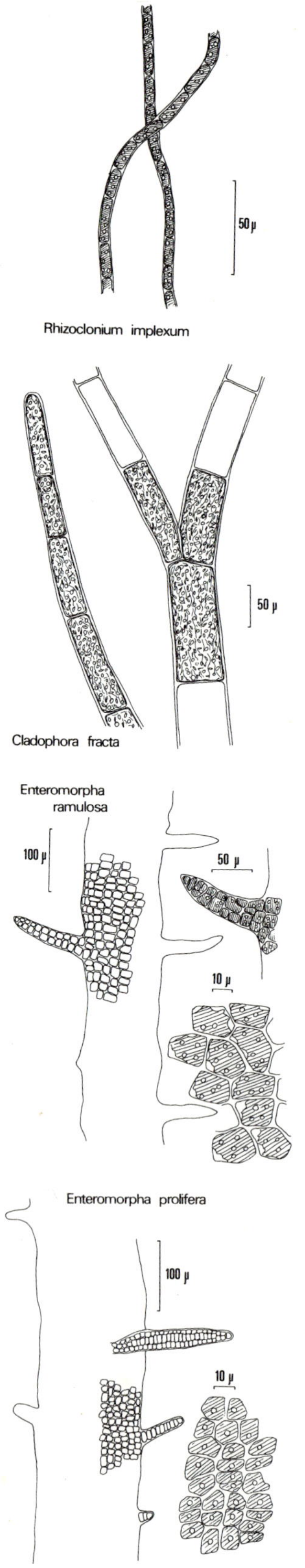

Fig. 4:3. Four of the green algae abundant on the mudflats. The pyrenoids and extent of the chloroplast are shown in the larger-scale drawings. *Rhizoclonium implexum* and *Cladophora fracta* formed the bulk of the 'algal mat' north of the causeway in mid-Summer 1975. They were partly replaced by *Enteromorpha ramulosa* and *E. prolifera* in the Autumn.

most numerous of these are teal, shelduck and pintail which all reach their maximum between November and January *(Fig. 8:1)*. In addition to alga, duck are said to eat *Salicornia,* but by the end of November there is very little *Salicornia* left and alga is probably their sole food. Alga is also eaten by brent geese, which are said to prefer *Zostera,* but in mid-winter it is quite usual to see geese and find their droppings well away from the main *Zostera* beds. On examination, the droppings are seen to contain little besides the remains of *Enteromorpha* plants. In February and March very little alga was collected along the transect and in many places the mud flats were bare. During these months most of the wildfowl leave the area and it seems likely that they do so because food is scarce.

In April the air temperature rose *(Fig. 4:2B)* and alga began to grow again between the Santry River channel and the *Salicornia* beds. Most of the brent geese depart in April. By May alga from the area close to the shore had begun to drift away and no doubt contributed to the biomass recorded at other sites; re-growth had begun between the *Salicornia* area and the salt-marsh proper. During June and July there was a steep increase in the amount of alga over the whole area, except in the *Salicornia* beds. Between the *Salicornia* beds and the edge of the salt-march cliff there was a marked decrease in biomass recorded at the end of August. Inherent senescence, perhaps hastened by self shading in the dense growth of *Rhizoclonium arenosum* and *Cladophora fracta,* appeared to be the cause of this.

Temperatures during June, July and August 1975 were considerably above the average for 1968–72. This may have caused the growth of the algae during this period to differ slightly from the normal pattern.

Accompanying the seasonal changes in algal biomass there were changes in the composition of the algal communities. In November and December the predominant species across the mud flats were

Enteromorpha prolifera, E. ramulosa (Fig. 4:3) and *E. flexuosa. E. ralfsii* occurred between the Santry river channel and the *Salicornia* beds. On the *Salicornia* beds much drifted material clung to the skeletons of the plants as the tide receded. *Ulva lactuca* was quite abundant in this area and *Enteromorpha torta* was also recorded. The following July and August the most abundant alga overall was *Rhizoclonium arenosum.* This was present in great quantity, mixed with *Enteromorpha ramulosa* and *E. flexuosa,* between the main channel and the *Salicornia* beds. Between the *Salicornia* and the salt-marsh cliff the latter species were only present in small amounts and *Rhizoclonium arenosum* was mixed with *Cladophora fracta. Ectocarpus silicuosus* and narrow filaments corresponding to descriptions of *Rhizoclonium kochianum* Kutzing were found along the length of the transect. *Enteromorpha prolifera* was scarce and only recorded close to the mainland coast. During May and June the mud surface between *Salicornia* stems at two stations near the middle of the transect was covered with a matted growth of *Rhizoclonium implexum (Fig. 4:3).* This had almost completely disappeared in August, as had a similar mat of *Blidingia marginata* growing among *Salicornia* just below the salt-march cliff. The August sample for this last station consisted of drifted material of *Ulva lactuca* and *Enteromorpha spp.*

These observations on the seasonal changes in the species composition of the algal communities suggest that the algae differ in the length and timing of their growth cycles. *Enteromorpha prolifera, E. ramulosa* and *E. flexuosa* persist throughout the year, though in the part of the mud-flats I have studied closely, appear to reach a peak in the Autumn. It seems likely that this is caused by a period of Autumn growth of these species or by material being carried by the tides from other parts of the flats during the Autumn. *Rhizoclonium implexum, Rhizoclonium arenosum, Cladophora fracta (Fig. 4:3)* and *Blidingia marginata* all grow rapidly in the early part of the summer and reach a peak in July. Their growth cycle is relatively short, especially so in the case of *Blidingia marginata* and *Rhizoclonium implexum.*

It is interesting to compare this with the observations of Chapman (1939) for the mudflats at Scolt Head Island in Norfolk. He describes three types of algal community: a 'lower sandy chlorophyceae consocies', in which *Enteromorpha prolifera* is dominant and *E. compressa* and *E. intestinalis* occur with *Vaucheria spp.;* a 'sandy chlorophyceae consocies' in which *Entermorpha torta* and *Rhizoclonium implexum* were dominant and *Enteromorpha compressa, E. prolifera, E. ramulosa, Rhizoclonium arenosum* and *Blidingia minima* occur; and a 'muddy chlorophyceae consocies' in which *Enteromorpha prolifera* and *Marostroma grevilliae* are dominant and *Cladophora fracta, Enteromorpha torta, Ulva lactuca* and *Ectocarpus silicuosus* occur. Both of the sandy cholorophyceae associations attained their maximum development during the summer, but the muddy association was at its maximum between January and May.

The larger brown algae

An account of the algae of the mudflats would be incomplete without reference to the diversity of the Fucoids in the intertidal area north of the causeway. Fragments of *Fucus vesiculosus* and *F. spiralis* are often found on the mudflats at low tide. These are peculiar in that they have numerous proliferations from the base of the fronds. In some drifting fronds which become buried by sediment the proliferations continue to grow up through the mud, forming small tussocks. Normal plants of *F. vesiculosus (Plate 6:4)* and *F. spiralis (Plate 6:3)* grow on the permanent concrete structures which extend into the mud from the sea wall at Kilbarrack. These two species also grow on stones in the mud on this part of the shore. In this habitat they are accompanied by many plants which are intermediate in their characters between *F. spiralis* and *F. vesiculosus.* A population with many intermediates has been described from the River Mersey (Burrows & Lodge 1951) where the plants were widely spaced, as they are on stones in the mud at Kilbarrack. Burrows & Lodge suggest that the Mersey population is a hybrid swarm, and that the absence of competition in an open population has allowed the barriers between the species to be broken down. This may well be happening to *Fucus spiralis* and *F. vesiculosus* on the Kilbarrack flats. *Fucus serratus* occurs in large isolated clumps close to the edge of the permanent channel. To the north west of the island there are no intermediates between it and either of the other species.

Chapter 5
THE FAUNA OF INTERTIDAL FLATS AND BEACH

A. B. West *Zoology Department Trinity College*

Introduction

The intertidal flats, like other littoral habitats, are submerged twice daily by the rising tide and exposed as often so that the animals which live there have a sort of Jekyll and Hyde existence, and most are active only when the tide is in. But the intertidal flats differ from other littoral habitats in several important respects. Here the substrate is devoid of solid supports for animals such as mussels and barnacles to attach themselves, and equally lacking are the crevices, sheltered overhangs and protected nooks associated with such supports. The flats provide only one form of shelter for their fauna: they allow animals to burrow. Those which do not do this are ready targets for predation by the birds which feed on the flats when the tide is out and by the sublittoral animals, chiefly fish, which do so when it is in. Consequently about 85% of species found on the Bull Island flats are infaunal (i.e. burrowers).

All intertidal habitats are susceptible to sudden changes in salinity, for example when the tide falls during rain. For the infauna of mud and sand this poses virtually no problem since the fresh water flows off the surface and hardly penetrates at all. However, the flats are susceptible to salinity changes of a much more pervasive kind due to their location in an estuary. When the rivers are in high flood the salinity of the water covering the flats at high tide may be significantly lowered. This dilute sea-water cannot be avoided by most of the fauna although a few animals, such as the snail *Littorina saxatilis* and the cockle *Cardium edule*, can seal off their shells tightly and escape it for a while. Consequently the majority of animals on the flats are euryhaline (i.e. can tolerate fluctuations in salinity) and some, e.g. *Nereis diversicolor,* the ragworm, can even live for long periods in fresh water.

The Bull Island flats, as has been explained in detail in *Chapter 2*, are depositing shores on which sedimentation is taking place continuously. The particles which are being deposited include large quantities of organic material, chiefly detritus (i.e. dead particulate organic material of diverse origin). This, by virtue both of its own composition and of the abundant micro-flora which grows on it, is a protein and energy-rich nutrient source for the fauna. Thus, although the fauna of the intertidal flats is restricted mainly to infaunal euryhaline organisms and consequently is composed of a rather small number of species, the numbers of individuals are enormous and the total biomass is relatively very great. This substantial food supply is one of the factors which enables the huge bird population to overwinter on Bull Island.

The intertidal continuum

The intertidal flats form a continuum around North Bull Island which may be divided into four sub-environments, *Salicornia* flats (*Plate X*), mudflat (*Plate XIII*), sandflat (*Plate VI*) and beach (*Plate II*). Each can be easily recognised and described *(Chapter 2)* although the boundaries, situated on gradients, are impossible to draw.

The mean particle size and organic content of the substrates each have a strong influence on the composition of the fauna. The former increases and the latter decreases from *Salicornia* flat to mud to sandflat to beach (*Fig. 5:1*). Where there is low mean particle size and high organic content the sedimentation rate is high and there is plenty of food: thus the mudflat supports a larger biomass than the beach. (The *Salicornia* flat is a special case and will be discussed later). Muds, possessing finer particles tend to be more thixotropic than sands which means that, for a burrower, not only is the fine substrate easier to penetrate but also permanent burrows can be maintained more easily in its stable matrix.

On the beach oxygen penetrates to a good depth in the sand as is evident from the depth to the black anaerobic layer. Animals living infaunally here have no problems of obtaining oxygen. In mud, however, where the organic content is higher, the oxygenated layer is often only a few millimetres deep and most of the infauna live in the anaerobic zone. These organisms overcome the difficulties of oxygen deprivation either by possessing physiological mechanisms which enable them to respire anaerobically, e.g. *Tubifex costatus,* or by maintaining a pipe-line to the surface through which they obtain their oxygen supply. For example, the latter strategy is adopted by *Corophium* (*Plate 5:1*) and the lugworm *Arenicola* (*Plate 5:2*), whose oxygenated burrow walls stand out brown against the black mud through which they run; and by bivalves such as *Macoma* and *Scrobicularia.*

The most common members of the macro-fauna of the intertidal flats, excluding birds, are listed in *Table 5:1* with indications of their habitats, highest recorded densities on Bull Island, feeding habits, etc. Their distribution on the flats is given in *Table 5:2.*

The mudflat substrate has a small mean particle size, high organic content, shallow oxygenated layer and is very thixotropic. The infauna includes a large proportion of species with permanent or long-lasting burrows, e.g. *Nereis diversicolor, Mya arenaria* (*Plate 5:3*), and *Corophium volutator,* reflecting the stability of this medium. In feeding habits the detritivores predominate (as in the other sub-environments, *Table 5:3*), and there is a high proportion (38%) of substrate ingesters, e.g. *Peloscolex benedeni, Scoloplos armiger.* That this is also a good environment for suspension feeding is reflected in the 12% of species that engage in this habit. The mudflats are, as it were, farthest up the estuary and nearest to the outflow of fresh water (in fact, each lagoon may be regarded as an estuary in its own right, with a small river entering on either side of the causeway) and have a euryhaline fauna. The oligochaetes *Peloscolex* and *Tubifex,* the polychaetes *Arenicola, Capitella* and *Nereis,* and the molluscs *Hydrobia ulvae, Littorina saxatilis, Cardium, Scrobicularia* and *Mya* are notable in this respect. To the naturalist on the shore, however, by far the most conspicuous animal is the little gasteropod *Hydrobia ulvae* which occurs on and within the mud in numbers up to 25,000 per square metre. (See also *Plate X.*) The biomass here is undoubtedly great, and although there are no satisfactory figures for comparison with the other sub-environments it is probably safe to say that it is greater than on the sandflats or beach.

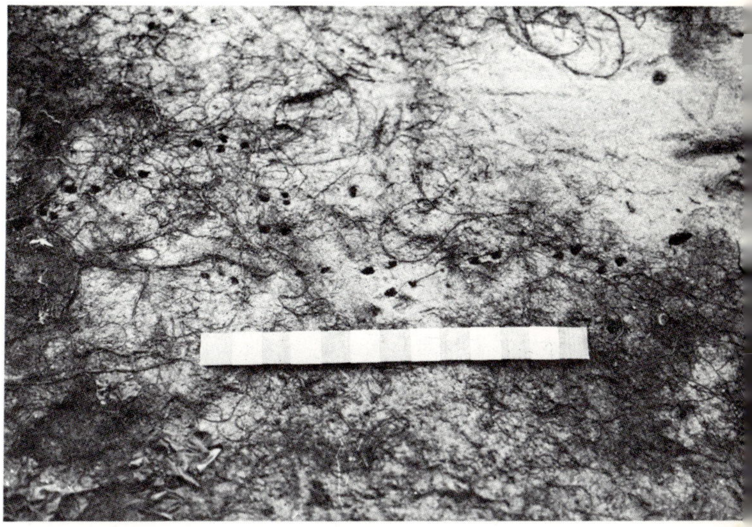

Plate 5:1. Burrows of *Corophium volutator.* Scale in cm.

Plate 5:2. *Arenicola* burrows, head shaft and tail cast.

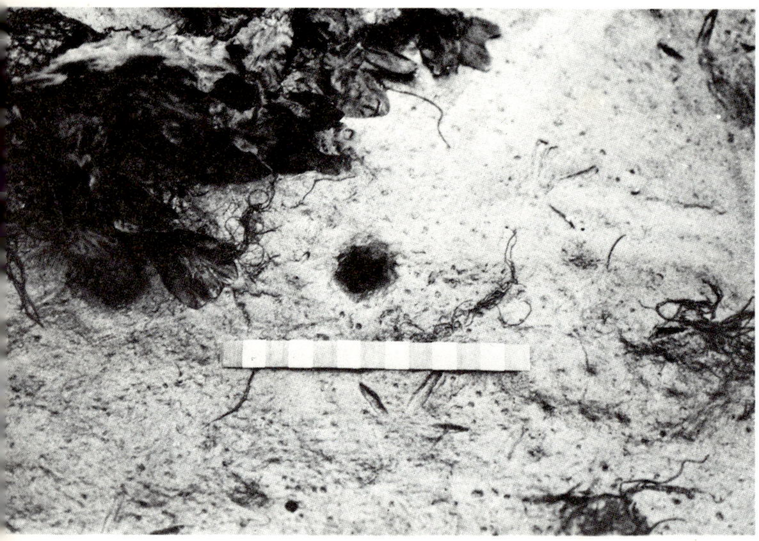

Plate 5:3. A *Mya* burrow in the mussel bed area with *Fucus spiralis.*

Plate 5:4. *Mytilus edulis* in the mussel bed off Kilbarrack. Note the small barnacles on some of the mussel shells. The brown algae *Fucus vesiculosus* and *Porphyra umbilicalis* are attached to this firm substrate. Scale in cm.

The sandflats are intermediate in characteristics between the mudflats and the beach. One of their most conspicuous features is the abundance of *Arenicola* casts which litter the surface (*Plate 5:1*). The presence of *Arenicola* and *Nereis* indicate that the substrate is stable enough to support permanent burrows, though these species are absent from some of the sandier areas. Here some other tubicolous polychaetes are found, *Spio filicornis* and *Lanice conchilega*, which make their tubes by cementing sand grains together with mucus. In comparison with the mudflats there are fewer substrate ingesters and more deposit feeders among the sandflat species. This reflects the greater efficiency of selective deposit feeding in extracting nutriment from substrates containing a lower percentage of organic material.

The beach, like the other sub-environments, shows continuous gradation. Its sandiest part is towards the east and it becomes quite muddy towards the Bull Wall. It has the highest mean particle size and lowest organic content of the sub-environments, and it is oxygenated to the greatest depth. It is a relatively mobile and unstable substrate, being churned up at the surface by wave action. The few species which have permanent burrows, e.g. *Lanice* and *Spio* make tough tubes which resist abrasion. The majority of the infauna have no permanent burrows but can bury themselves in the substrate, e.g. the amphipods *Bathyporeia, Urothoe, Haustorius,* or move through it, e.g. *Scoloplos, Tellina.* The amphipods, unlike *Corophium* on the mudflats, need no pipeline or oxygen and maintain no link with the surface. These amphipods, and the isopod *Eurydice,* are species characteristic of the beach and are found in none of the other sub-environments. They are all deposit feeders, and contribute to the high percentage of species of this habit (60%) found here. Another abundant species typical of this habitat is the bivalve *Tellina,* whose wafer-thin streamlined shell and knife-like foot are adaptations for movement through a dilatant substrate. The beach is the least estuarine of the sub-environments, and its fauna is the least euryhaline. It appears to support the smallest biomass.

The *Salicornia* flats *(Plate XI)* are physically very similar to the mudflats which surround them, but the organic content is slightly higher. Their fauna has not received much attention in the past, but we do know that the density of oligochaetes here is much greater than in the surrounding mudflat *(Table 5:2).* The increase in numbers is primarily due to species other than *Peloscolex benedeni* which is the dominant species on the mudflats. *Peloscolex* itself, however, is also more numerous here than on mud and densities of up to 75,000 per m^2 have been recorded. Substrate ingesters, predictably, are the dominant feeding type *(Table 5:3)* and deposit feeders are common. The high position of the

Salicornia flats on the shore probably makes it difficult for suspension feeders to survive here as their feeding time would be very restricted. It may also be an important factor in determining which oligochaetes occur since there is a strong similarity with the oligochaete fauna of the *Puccinellia* zone of the saltmarsh.

The algal mat *(Chapter 4)* is a feature of sand, mud and *Salicornia* flats which influences their faunas in a number of different ways. The mat traps fine suspended particles and the underlying substrate becomes rich in organic matter, and thus completely anaerobic. Substrate ingesters tolerant of oxygen lack, such as *Peloscolex* and *Capitella,* are abundant. Within the mat the gastropod *Hydrobia (Plate X)* is often extremely numerous, and *Littorina saxatilis* may be common. It is one of the few sources of shelter on the flats and young crabs, *Carcinus,* commonly hide in it.

Considering the relatively great biomass of the algal mat, at least at certain times of the year, it is at first surprising that no members of the invertebrate fauna graze on it *(Table 5:3)*. This is said assuming that *Hydrobia* and *Littorina saxatilis* do not feed directly on the algae, although there is slight doubt about this. Marine littoral macro-algae in general are grazed only by a few invertebrates, none of which are common on Bull Island flats. Feeding on the algal decomposition products seems to be a more widespread habit. This has the advantage that it overcomes the seasonal availability of the algal mass, which presents no problems for the grazing birds since their presence on the flats is linked to it.

Three specialised habitats

Three sub-environments fall outside the major categories of the continuum discussed above: the channels which traverse the flats, the mussel banks, and the interface between the flats and the terrestrial environment. Their faunas differ from those of the neighbouring major sub-environments, but they are poorly studied and can only be described briefly.

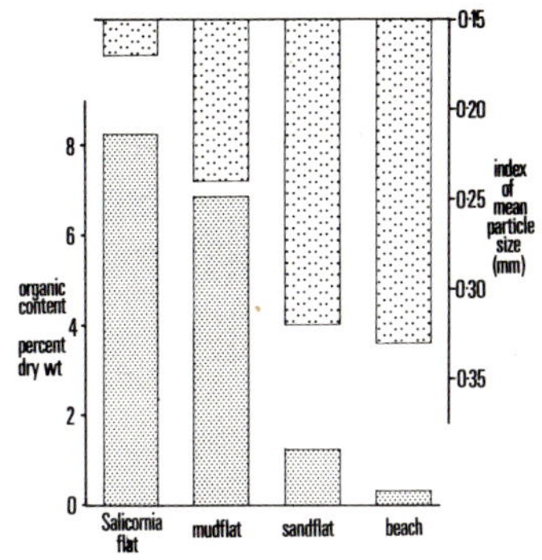

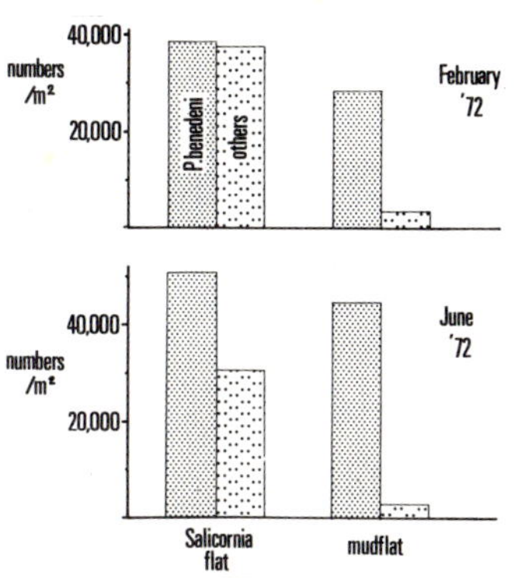

Fig. 5:1. Organic content (% dry weight) and index of particle size (reciprocal of the mean of mean phi value) of intertidal substrates. Data derived from Harris (1973).

Fig. 5:2. The mean numbers of *Peloscolex benedeni* and other oligochaetes from samples taken from Salicornia flat and mudflat. Data derived from McGrath (1972).

Channels

The channels carry fresh water run-off across the flats when the tide is out. At high tide they carry sea water into some parts of the terrestrial region, e.g. in the salt marsh where the pans and channels are to some degree extensions of the mudflats *(Plate XII)*. The channels then are subject to great fluctuations in salinity, and all the species which occur in them are euryhaline *(Table 5:2)*. Infauna includes the polychaetes *Nereis* and *Arenicola,* the bivalve *Cardium,* and, towards the edges of the channels, the amphipod *Corophium. Hydrobia* and *L. saxatilis* are often abundant. The presence of permanent water makes it possible for swimming animals to occur here and among these the estuarine goby *Pomatoschistus* and the shrimp *Crangon,* both of which can bury themselves in sand or mud, are probably the most common. The occurrence of *Pomatoschistus* is seasonal, as is described in *Chapter 12*.

Table 5:1. A brief directory of some of the most common animals of the intertidal flats.

Species	Habitat (+ maximum nos/m²)	Feeding	Notes
Phylum Annelida – Class Oligochaeta			
Peloscolex benedeni	Mud, often anaerobic (75,000)	Substrate ingestion	Brown worm, about 1 cm by 0.5 mm. Lives in the top 2 cm of mud. Tolerates anaerobic conditions. Occurs in great numbers under algal mat and in *Salicornia* flat.
Tubifex costatus	Mud, often anaerobic	Substrate ingestion	Similar in size to *Peloscolex*. Tolerates anaerobic conditions. Abundant in *Salicornia* flat.
Phylum Annelida – Class Polychaeta			
Arenicola marina (lugworm)	Muddy sand and sandy mud (400)	Deposit feeder: non-selective substrate ingestion	Large worm, often 20 cm by 1 cm. Has U-shaped burrow, lined with mucus. Ingests mud and sand in the headshaft of the burrow, causing depression at the surface. Defaecates the characteristic cast at the top of the tail shaft. Pumps water through the burrow to provide oxygen, and hence the brown walls of the burrow running through black mud. Young worms occur on the shore.
Capitella capitata	Mud, usually anaerobic	Deposit feeder	Red, 3–10 cm by 1 mm. Tolerates low oxygen tensions. Common under algal mat.
Lanice conchilega (sandmason)	Clean sand	Deposit feeder. When submerged, tentacles extend from top of tube and collect food particles from sand surface	About 10 cm by 5 mm. Inhabits a straight tube, up to 30 cm long, of mucus encrusted with sand. Upper 2–5 cm stick up like a sand-covered cigarette with a frayed top. Abundant in places, e.g. on steeply sloping shore at north side of recurve.
Nephthys hombergi (catworm)	Sand, sandy mud	Carnivore, preys on other infauna	About 5–10 cm long by 3 mm. Has no permanent burrow (errant form).
Nereis diversicolor (ragworm)	Mud, sandy mud (1,500)	Carnivore, but can also be suspension or deposit feeder	Up to 15 cm by 5 mm. Inhabits permanent ramifying burrow with a single opening at the surface, but often 30 cm deep. Extremely euryhaline, surviving even in fresh water although it cannot reproduce in it.
Pygospio elegans	Mud, muddy sand	Selective deposit feeder	About 1 cm by 1 mm. Inhabits mucus-lined burrow.
Scoloplos armiger	Sand, muddy sand	Non-selective deposit feeder, like *Arenicola*	Up to 10 cm by 4 mm. Has no permanent burrow. Lays eggs in characteristic balloon-shaped gelatinous masses, about the size of a marble, anchored to the sand by a mucous string.
Spio filicornis	Mud, muddy sand	Selective deposit feeder	About 1 cm by 1 mm. Inhabits sand-encrusted mucous tube of 1–2 mm diameter and up to 5 cm long.

Phylum Arthropoda – Class Crustacea			
Carcinus maenas (shore crab)	Most shore types	Scavenger and predator	Numerous where there are stones or rocks. Young crabs live on the flats under cover of the algal mat.
Corophium volutator	Mud, sandy mud (50,000)	In burrow, irrigatory current is used for suspension feeding. Also forages for detritus	Elongate white amphipod about 1 cm long, 1 mm thick. Inhabits a U-shaped burrow, about 5 cm deep, with two D-shaped apertures about 1 cm apart at the surface. Probably most abundant animal on mud and sand flats.
Bathyporeia pelagica *B. robertsoni* *Haustorius arenarius* *Urothoe marina*	Clean sand	Feed on detritus in the sand	These amphipods burrow into sand. They emerge and swim actively when the tide is in, especially by night.
Eurydice pulchra	Clean sand	Feed on detritus in the surf	This isopod is found high on the sandy beach. It lies buried when the tide is out but emerges when waves break over it.
Phylum Mollusca – Class Gastropoda			
Hydrobia ulvae	Mud, sandy mud, algal mat (25,000)	Deposit feeder	Usually less than 7 mm long, this snail occurs on the surface and buried to a depth of 1 cm. It harbours the larval stages of a number of trematodes which are adult in waders and waterfowl, for which it is an important food item.
Littorina saxatilis (rough winkle)	Rock, mud, boulder, shingle	Deposit feeder	Less than 1.5 cm long, but much more robust than *Hydrobia*. Very widespread on different shore types. Euryhaline and tolerant of severe desiccation. Probably spreads from Bull Wall, saltmarsh and stony parts of the shore onto the mudflat.
Phylum Mollusca – Class Lamellibranchia			
Cardium edule (cockle)	Sand, muddy sand	Suspension feeder (see notes)	Bivalves collect food by pumping water through an inhalent siphon to a filter, which removes food particles, and out via an exhalent siphon. In *Cardium* the siphons are short and fused. It lives on or just below the surface, and extends its siphons to clear the substrate when feeding.
Macoma balthica (baltic tellin)	Mud, sandy mud	Siphons separate. Inhalent sucks in deposits from surface or particles in suspension	A mobile species, usually found within 6 cm of the surface.
Mya arenaria (sand gaper)	Fine mud	Suspension feeder	Often found 30 cm deep in mud, this bivalve stays always in the same position. The long siphons are fused together forming a thick tube which cannot be withdrawn into the shell.
Mytilus edulis (mussel)	Stones and other solid supports	Suspension feeder. Siphons similar to *Cardium*	Found wherever suitable supports occur on which it can grow. Unlike the other bivalves mentioned it attaches to a support by strong threads, and remains there. Euryhaline.
Scrobicularia plana	Mud	As *Macoma*	Similar to *Macoma,* but larger, deeper-dwelling and less common.
Tellina tenuis (thin tellin)	Clean sand (200)	As *Macoma*	Characteristic of sand. The wafer-thin shell and knife-like foot enable it to move through dilatant substrates.
Phylum Chordata – Class Pisces			
Pomatoschistus microps (estuarine goby)	Permanent brackish water	Carnivore	This euryhaline goby usually lies partially buried in sand or mud. It is found onshore in autumn, winter and spring but migrates offshore to breed in summer.

Mussel banks

The mussel banks border the main channel of the Lagoon east of the Causeway (*Plate 5:4*). Mussels, unlike the other bivalves of the flats, require solid substrates to which they attach permanently by means of their byssus threads. In the mussel banks clumps of these bivalves adhere to the shells of their dead predecessors and also to one another. Towards the east end of the channel the banks support a rich fauna of invertebrates including sponges, hydroids and anemones. Towards the west, however, the banks are becoming progressively silted up by the rapid sedimentary processes which are taking place near the Causeway, and they are doomed to extinction. Here the mussels are surrounded by schlick mud, and infaunal animals such as *Nereis*, *Mya* and *Scrobicularia* predominate. The mussel banks, if mapped in successive years, might provide an interesting index of silt deposition, contracting as the *Salicornia* flats extend.

Littoral – terrestrial interface

The littoral-terrestrial interface of the lagoon is a category of convenience and not a uniform habitat. It includes those regions of the shore which lie below high watermark and above one of the four major sub-environments. For example, the mainland shore between the Bull Bridge and the Causeway has a zone of stones and small boulders above the level of the flat. These provide solid surfaces to which barnacles *Balanus balanoides*, mussels *Mytilus edulis*, and fucoids (brown algae) can attach and on which the gastropods *Littorina littorea*, *L. saxatilis* and *Hydrobia* congregate. The fucoids form a suitable habitat for the flat winkle *L. littoralis*, which grazes on them. *Carcinus maenas* the shore crab occurs under the shelter of stones and fucoids, and two isopods also are found in sheltered situations: *Jaera albifrons* near sources of fresh water and *Ligia oceanica* in crevices high on the shore.

Another part of the interface which warrants investigation is the area between the saltmarsh and the flats (*Plate 1*). Here an abrupt change in the oligochaete fauna has been detected. Within three metres of the edge of the marsh a fauna dominated by terrestrial species changes to one chiefly of littoral species. In this region two arthropods are frequently met with, the hopping amphipod *Orchestia gammarella* and a tiny blue wingless insect, the collembolan *Anurida maritima*.

	Salicornia flats	Mudflats	Sand flats	Beach
Algal grazers	0	0	0	0
Detritivores:				
Substrate ingesters	50	38	20	13
Deposit feeders (D)	33	19	33	60
Suspension feeders (F)	0	12	7	7
Optional D or F	17	12	13	7
Total detritivores	100	81	73	87
Omnivores	0	6	7	0
Carnivores	0	12	20	13

Table 5:3. Numbers of species in different feeding categories, expressed as a percentage of the total number of classifiable species in each habitat.

A comment on our knowledge of the fauna

It is not possible here to treat either the microscopic interstitial fauna of the flats, although both foraminifera (protozoa) and harpacticoidea (crustacea) have been investigated, or the temporary macrofauna. These visitors include rats which visit the shore by night, hares which have been reported making forms in banks on the mudflats, jellyfish which are stranded on the beach, and the other visitors which come in with the tide – for none of which there exist many records.

The information upon which this account is based is derived mainly from unpublished data which has been made available through the kindness of those who collected it. This data is not weighty. It has been spun together here in an attempt to synthesise an account of the fauna, elucidating its role in this complex ecosystem. The data alone, however, was not sufficient for this, and here and there the cements of experience from elsewhere and of "informed speculation" have been liberally applied. No reader should go away, because of the picture painted here, with the impression that the fauna of the intertidal flats is well understood or adequately studied. On the contrary, there are immense gaps in our knowledge of it. Many of these could be filled by the rigorous and rewarding investigations of amateur naturalists, like those which have contributed so significantly to our knowledge of the birdlife of the area. It is to be hoped that some will take up this fascinating challenge.

The author wishes to thank the following workers who have made unpublished material available for use in this chapter. V. Canning, A. E. Elliot, J. N. R. Grainger, B. Healy.

	Salicornia flats	Mudflats	Sandflats	Beach	Channels	Littoral-Terrestrial Interface
Annelida – Oligochaeta						
Peloscolex benedeni	★★★★	★★★ (★★★★)	(★★★)	.	★	.
Tubifex costatus	★★★	★	.	.	.	.
Enchytraeids	★★	★	.	.	.	.
Annelida – Polychaeta: Sedentaria						
Arenicola marina	.	★	★★★★	★★	★★	.
Capitella capitata	.	(★)	.	.	.	.
Spio filicornis	.	.	★★	★★	.	.
Pygospio elegans	.	.	★	★	.	.
Lanice conchilega	.	.	★	★	.	.
Annelida – Polychaeta: Errantia						
Nereis diversicolor	.	★★★	★★	.	★★	.
Nephthys hombergi	.	★	★★★	★★	.	.
Scoloplos armiger	.	★	★★★	★★	.	.
Eteone longa	.	.	★	.	.	.
Nephthys caeca	.	.	.	★★	.	.
Mollusca – Gastropoda						
Hydrobia ulvae	★★★★	★★★★	(★★★)	.	★★	★★★
Littorina saxatilis	★★	★	(★)	.	★★	★★★
Littorina littorea	.	.	.	.	.	★
Littorina littoralis	.	.	.	.	.	★
Mollusca – Lamellibranchia						
Cardium edule	.	★★	★★★	★★	★	.
Macoma balthica	.	★★	★★	★	.	.
Scrobicularia plana	.	★	.	.	.	.
Mya arenaria	.	★	.	.	.	.
Tellina tenuis	.	.	.	★★★	.	.
Crustacea – Ostracoda						
Leptocythere lacertosa	.	★★★	★★★	.	.	.
Leptocythere pellucida	.	★★	★★	.	.	.
Semicytherura sella	.	.	★	.	.	.
Cytherois fischeri	.	.	★	.	.	.
Pontocythere elongata	.	.	.	★	.	.
Crustacea – Isopoda						
Eurydice pulchra	.	.	.	★★	.	.
Jaera albifrons	.	.	.	.	.	★
Crustacea – Amphipoda						
Corophium volutator	★★★★	★★★★	★★	.	★	★★
Bathyporeia pelagica	.	.	.	★★	.	.
Bathyporeia robertsoni	.	.	.	★★	.	.
Urothoe marina	.	.	.	★★	.	.
Haustorius arenarius	.	.	.	★	.	.
Orchestia gammarella	.	.	.	.	.	★
Crustacea – Decapoda						
Crangon vulgaris	.	.	.	★	★	.
Carcinus maenas	.	(★)	(★)	.	★	★
Chordata – Pisces						
Pomatoschistus microps	.	.	.	.	★★★	.
Anguilla anguilla	.	.	.	.	★	.
Mugil labrosus	.	.	.	.	★	.

Table 5:2. The relative abundance of some common animals in different habitats on the intertidal flats. Data in parentheses refer to abundance in the algal mat.

Chapter 6

INTERTIDAL FLATS AS ECOSYSTEMS

D. W. Jeffrey *Department of Botany* *Trinity College*

Ecosystems in general

The ecosystem concept is a unifying idea that may be usefully applied to complex ecological situations. The intertidal flats are used to illustrate this approach, making use of the extensive data available on feeding. It must be remembered, however, that this account is based as much on material from the literature as current local work. No field study of any Bull ecosystem has yet been attempted, and this exercise could have been legitimately applied to sand dune or saltmarsh ecosystems.

An ecosystem may be defined as a more or less self-contained biological system in which energy for biological processes is transmitted from one class of organism to another and within which various substances are circulated. Energy, of course, originates from the sun and is absorbed in photosynthesis. A simple block diagram (*Fig 6:1*) permits us to visualise the idea of energy transfer more clearly. The boxes represent the actual quantities of fixed carbon energy, which are closely reflected by biomass of each type of organism. It will be realised that in such a food chain the biomass of green plants supports a lesser biomass of herbivores and a still smaller carnivore biomass. Ten-fold reductions in weight or number of organisms are commonly recorded between each class. The reduction results from both respiratory losses and transfers to the variable pool of non-living detritus. Detritus is the energy source for the decomposer class, fungi, bacteria, protozoa and small invertebrates, which mop up remaining energy.

Highly efficient transfer of all manner of vital substances occurs together with energy. Nitrogen, for example, is absorbed by green plants as ammonium or nitrate and assimilated into amino acids and proteins. Subsequent nitrogen transfer occurs as largely insoluble protein. Ions such as phosphate, sulphate, potassium, calcium and iron are effectively transferred and often accumulated, along the food chain. The sheer effectiveness of transfer and accumulation leads to the vulnerability of ecosystems to poisoning by D.D.T., toxic metals such as mercury, lead and copper or radio isotopes such as strontium-90. Such poisons tend to attain high concentrations in animals at the top of the food chain.

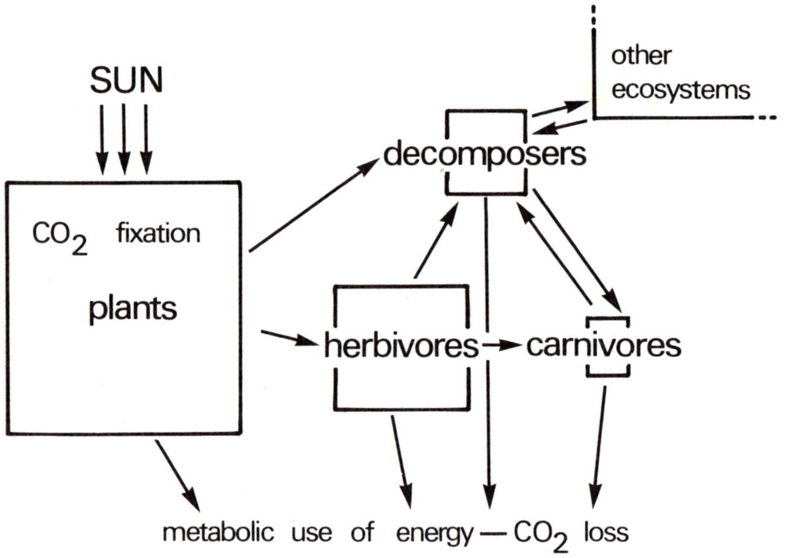

Fig. 6:1. Diagram to illustrate the pathways for energy transfer in a typical ecosystem. Input from other ecosystems and the role of decomposer organisms is exaggerated in estuarine situations.

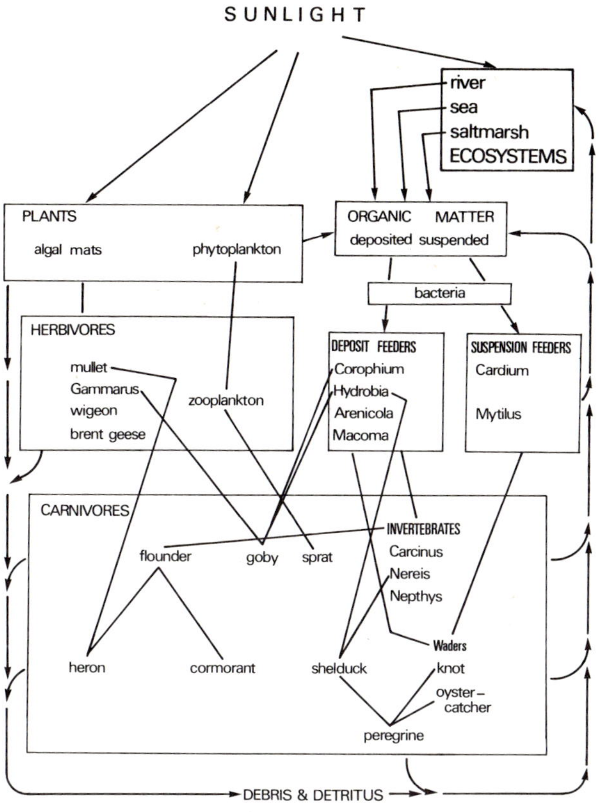

Fig. 6:2. Energy flow diagram for the Bull Island intertidal flats, based on information from chapters 4, 5 and 9. Only the adult forms of the most conspicuous organisms are considered, sometimes in generalised form.

Intertidal Ecosystems

How typical are intertidal ecosystems? The most casual glance at sand or mudflats makes one realise that here are ecosystems which differ from those most text books describe. Grasses and trees are absent, large herbivores or carnivores do not abound and the usual aerobic decomposer organisms are obviously inhibited in many areas. Furthermore, the sedimentary conditions of the estuary bring in large quantities of organic matter. This organic matter simply is interpreted as energy transferred from one ecosystem to another. It is probable that the amount of energy acquired in this fashion is at least as great as that absorbed in situ by photosynthesis. This exaggerated transfer places more emphasis on the decomposers, and leads me to depict two parallel food chains which ultimately interlock (*Fig. 6:2*). One is a classic food chain based on the algae and other plants of the intertidal flats. The other is based on the largely imported organic matter. Herbivores, on the one hand and detritivores on the other consume in various specialised ways their respective sources of energy. These animal groups are preyed on by a diverse set of predators, some very specialised, others having a wider range of food sources. At each of these levels of consumption, energy is lost by respiration, faecal materials are eliminated and organisms will die, recycling back to the prominent deposit and suspension feeders.

The arrows drawn on the diagram merely illustrate a known feeding or predator-prey relationship. It must be borne in mind that the ecological roles of invertebrates are often obscure, and it is not easy to determine, for example, the food or predators of a particular animal. Direct field observations of feeding behaviour, laboratory studies of feeding mechanisms and analysis of gut contents are primary means to this end. They are, incidentally, methods that may be within the scope of the amateur naturalist. The ecosystem will not be fully understood or quantified until more is known about the population dynamics, and seasonal cycles of both adult and larval forms of mud flat organisms. Very little is known about the way in which vital materials are distributed or transferred in this ecosystem. For example, it should be possible to draw up a nitrogen budget which would illustrate the flow of protein from plants and detritus to other organisms.

V. A similar situation to Plate I. but with marked erosion of the 'cliff' giving rise to a mud ball slope. The surface of the flat is densely covered with *Corophium* burrows. *(C. H. Holland)*

VI. A sandy sediment with burrows of *Corophium volutator* and shells of *Hydrobia ulvae.* Foot marks and probe marks of feeding birds are evident. The broad green algae is *Ulva lactuca* with a smaller quantity of an *Enteromorpha.* *(B. Healy)*

Interpretation

The ecosystem diagram is a basic and simplified account of the ecological relationships between a limited set of organisms. It may allow us to catch a glimpse of the forest, despite the complexities of each tree. Such an over-view is valuable to both the scientist seeking where best to place research effort and the conservation administrator. The latter needs to understand what he must do within existing legislation to carry out his duties in management.

Possible environmental impacts on the mudflat ecosystem

The Bull Island is successfully protected by law as a Bird Sanctuary, freedom from shooting being a significant factor in the conservation of birds. However, it will be seen that the few birds referred to in the ecosystem diagram are dependent on a diverse group of food sources. These range from herbivorous brent geese to the raptorial peregrine. The mudflat diversity must therefore not be eroded by declining environmental quality. There are many possible components to such erosion.

Complimentary to the food supply for the birds, is energy conservation. In winter months extremes of energy loss through radiated heat are additional to energy expenditure in flight and movement. Energy is conserved by roosting or remaining still, often with plumage inflated. An undisturbed environment is available at all states of the tide on the Bull, and provides safe roosting for birds feeding in all areas of the bay.

In a preliminary survey of the 10 km of shoreline from Clontarf to Sutton, approximately half of the shoreline could be classified as relatively unpolluted. Most of this unpolluted zone comprises the lagoonal mudflat areas *(Fig. 3:5)*, but a number of tendencies need to be continuously monitored. The accumulation of silt attributed to the causeway, has already been referred to. Its effects in ecosystem terms are to decrease the detritivore population to virtually a single species of oligochaeta. A drastic drop in diversity that is fortunately localised. Another ecosystem impact, related to the causeway is the high biomass of algae recorded between it and the Bull Bridge (*Fig. 4:1*). This is probably associated with the circulation of water with higher levels of nitrogen and phosphorus within the harbour area. This water is no longer diluted by mixing with the northern lagoon water and is thus absorbed by algal mats. These on the one hand increase food for herbivores, but decrease the areas of mud dwelling invertebrates.

In only one site is the level of a heavy metal, zinc, abnormally high – associated with the silt of the inner muds. This is probably another manifestation of urban pollution, arising merely from run-off from galvanised surfaces. Zinc is the least toxic of all the heavy metals, and here it is not even correlated with high lead or copper. However, these metals, plus cadmium and mercury may well be related to increasing industrialisation. They could be concentrated in a food chain such as we have described, and might produce a loss of large organisms.

Oil pollution is not so far a feature of the Dublin Bay environment. This is due to the relatively low tonnage of oil brought into the port, the good practice shown by the Port and Docks Board and a measure of good luck. Should either a major oil pollution event or repeated minor spills occur in the open Bay, the pattern of currents will ensure interaction with the intertidal flats. Various effects on this ecosystem may be predicted. Floating heavy oil has a devastating effect on feeding and roosting birds. This is dramatic and well recognised. Evidence is now building up that the light fraction of oils and the various detergents used to disperse oil have lethal effects on adult and larval invertebrates. Accumulation of emulsified oil has also been noted in suspension feeders, leading to toxicity in their predators.

In consideration of oil pollution, and other environmental impacts, the complexities of the intertidal flats are less well studied than the rocky shore. With the increasing industrialisation of estuarine sites there is a real practical need to improve our knowledge. Without understanding ecosystems, we cannot be rational about management for conservation.

Acknowledgement

It is appropriate to acknowledge at this point that an extensive survey of intertidal environment and organisms was sponsored by An Taisce – the National Trust. This study also contributed data to chapters 3, 4 and 5. Technical assistance in the survey was provided by David Baird, Peter Carvill, Grace O'Donovan, Rosemary Seymore and Jonathan Shackleton.

VII. Part of a flock of Brent geese resting in the middle marsh. The tide has covered their main feeding area but some are continuing to graze. The long-necked appearance indicates an awareness of the photographer. *(J. P. Newell)*

V C. H. Holland

VI B. Healy

VII J. P. Newell

VIII C. H. Holland

IX D. W. Jeffrey

X J. Shackleton

Chapter 7

A PALEONTOLOGIST'S VIEW OF SHELLY INVERTEBRATES

D. C. Palmer. *Trinity College*

Introduction

The presence of empty shells on a sandy beach is a common phenomenon within our everyday experience. Indeed to the casual observer life may not be much in evidence on the beach (*Plate 7:1*). There is rather more evidence of death with so many scattered remains of shelly molluscs.

A consideration of postmortem processes by which these remains are accumulated reveals a complex sequence of events, with the interaction of different biological and mechanical forces on the living fauna of a wide area. The study (called taphonomy) of these processes links biology and palaeontology.

The beach environment provides a useful laboratory for studying how living organisms can be transformed into fossils.

The aim has been to examine the living macrofauna of the beach, emphasising those organisms that are fossilisable. As such, they should have the greatest potential as recoverable evidence of life should this environment become a rock unit within the stratigraphic record of the future. In order to achieve this it is necessary to study both the living and dead fauna of the beach and the adjacent subtidal environment.

The animals common to the live and dead assemblages of any one locality can be determined with their relationship to their surrounding sediment. Thus for instance the attitude of the shells within the sediment indicates whether they lived and died in the same place or whether they were transported before final burial. From these and other observations it is possible to trace the postmortem development of the assemblage and to evaluate the extent to which these potential fossils reflect the known living fauna.

VIII. Lagoonal flat in Summer. The tendency of *Spartina anglica* to spread is illustrated, especially in this relatively muddy sand. Silt covered plants of *Salicornia* are in the foreground. *(C. H. Holland)*

IX. View of the N. E. lagoon from the causeway showing lower marsh with much *Puccinellia* and the widespread colonisation by *Salicornia* of the flats. Note the drift line at the high Spring tide mark. *(D. W. Jeffrey)*

X. *Salicornia dolichostachya* on the inner muds close to the causeway. The plants are growing in an area blanketed by *Enteromorpha* and typically supporting large numbers of *Hydrobia ulvae*. *(J. Shackleton)*

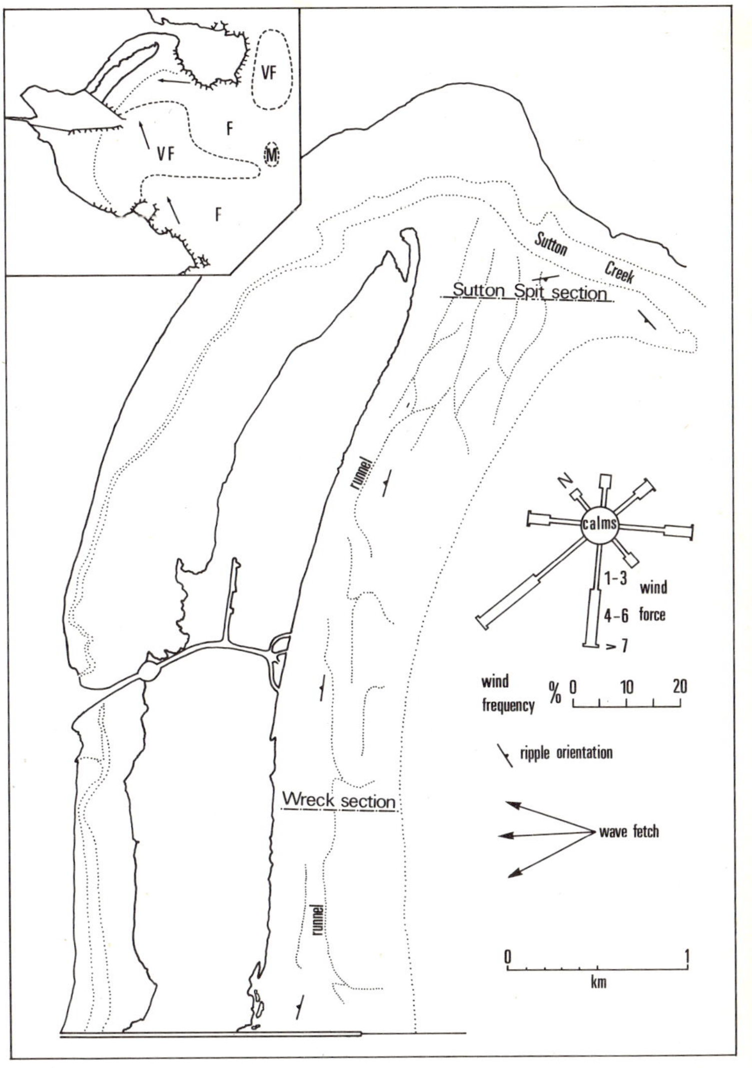

Fig. 7:1 (a) Outline map of Bull Island with location of sample sites along the two lines of section across the intertidal beach. Some morphological features of the foreshore are shown (runnels and ripple trends) with directional data for wave fetch and winds (including strength and frequency). (b) A map of sediment distribution in Dublin Bay with possible sources and derivation direction for the rocky foreshore epifaunas found within the dead assemblage of the sandy beach.

In a pilot study sampling was restricted to two transects (100 metres apart) down the beach slope (on average 1:250) normal to the shoreline (*Fig. 7:1*). At each sample site all the empty shells from a 50 cm² area of the sand surface were collected, as were both the live and dead faunas from 20 cm³ (approximately a spadeful) volume of sand. Sieving the latter separated the bulk of the shell material from the sand.

TAXA	number of species	% species	number of specimens	% specimens
Crustacea	8	38	201	67
Polychaeta	8	38	60	20
Mollusca	4	19	34	11
Nemertini	1	5	5	2
Totals	21	100	301	100

Table 7:1a. The taxa of the living fauna within the intertidal sandy beach (Sutton Spit section); their abundance and diversity.

Sutton Spit Section

species (locality)	1	2	3	4	5	6	7	8	9	10	total	%
Cardium edule	–	–	–	–	–	–	1	–	–	–	1	3
Macoma balthica	3	–	–	1	–	–	–	–	–	–	4	12
Tellina tenuis	–	–	6	11	4	2	4	1	–	1	29	85
											34	100

Wreck Section

species (locality)	1	2	3	4	5	6	7	*total*	%
Tellina fabula	–	–	–	1	–	–	1	2	5
Cardium edule	2	1	1	–	–	–	–	4	9
Donax vittatus	–	1	1	–	1	2	–	5	12
Tellina tenuis	6	6	4	8	7	–	1	32	74
								43	100

Table 7:1b. Distribution of live bivalves in beach transects at Sutton Spit and the Wreck. Sample sites 100 m apart.

The living fauna

An assessment of the total living fauna was only attempted for transect 1 (Sutton Spit, *Table 6:1*). It is apparent that both in terms of percentage of species and percentage of individuals representatives of the Crustacea and Polychaeta outnumber the Mollusca.

In the high energy sandy beach environment soft tissues are readily destroyed by mechanical abrasion. Consequently only those organisms with skeletal hard parts have a chance of preservation. Furthermore these hard parts have to be well mineralised and mechanically robust. The crustacea (as a whole) have exoskeletons and yet despite their abundance leave little in the way of remains in the sand because the skeletal tissue is not generally mineralised except in the crabs, lobsters, barnacles and ostracods. Although the polychaetes have no skeletons some genera do have preservable chitinous mouthparts *(Plate 7:2)* but they are a relatively minor group. Thus in the sandy beach environment the mollusca with their robust calcium carbonate shells dominate the dead assemblage even though they are not numerically dominant in the living assemblage. It follows that it is this group which has the greatest chance of being preserved for the fossil record. Consequently the record has severe limitations in the extent to which it represents the original living faunas of this environment.

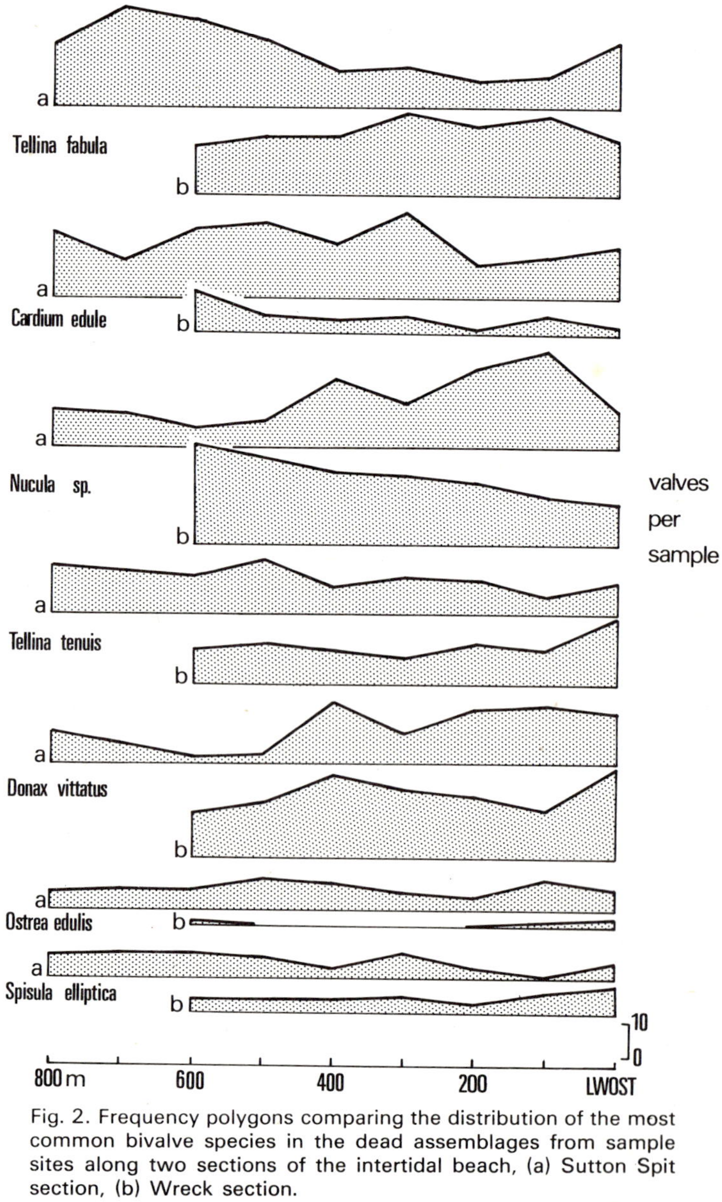

Fig. 2. Frequency polygons comparing the distribution of the most common bivalve species in the dead assemblages from sample sites along two sections of the intertidal beach, (a) Sutton Spit section, (b) Wreck section.

Table 7:2 a and b. The taxa of the dead assemblages from two sections of the intertidal beach *(a)* Sutton Spit section, *(b)* Wreck Section, with their species abundance relative to the total collections from each section.

Table 7:2a. Sutton Spit Section

	Bivalves	No. Shells	%	Cumulative
	Tellina fabula	353	18.9	18.9
Four species	Cardium edule	326	17.4	36.3
>10% each	Nucula sp.	228	12.2	48.5
	Tellina tenuis	219	11.7	60.2
	Donax vittatus	158	8.4	68.6
	Ostrea edulis	119	6.4	75.0
Six species	Spisula elliptica	106	5.7	80.7
>1% each	Mytilus edulis	47	2.5	83.2
	Chlamys varia	32	1.7	84.9
	Thyasira flexuosa	24	1.3	86.2
	Ensis siliqua	11		
	Venus striatula	11		
	Macoma balthica	5		
	Venus pullastra	5		
Nine species	Mactra corallina	3	2.3	88.5
<1% each	Cardium sp.	3		
	Dosinia lupinus	2		
	Gari fervensis	2		
	Corbula gibba	1		
	other bivalve specimens	52	2.8	91.3
	Gastropods:			
	Patella sp.	18	1.0	92.3
	Hydrobia sp.	14	1.0	93.3
	Littorina sp.	14	1.0	94.3
	Gibbula cineraria	7	0.4	94.7
	Natica sp.	1		
	Nassarius reticulatus	4		
	Diodora apertura	1	1.0	95.7
	Patina pellucida	1		
	Turbonilla sp.	1		
	other gastropod specimens	6		
	Echinoid fragments	17	0.9	96.6
	Bryozoan fragments	2	0.1	96.7
	Crustacea:			
	barnacle plates	73	3.9	100.5
	crab fragments	4	0.2	100.7
	total number of specimens	1,870		100.7

Table 7:2b. Wreck Section

	Bivalves	No. Shells	%	Cumulative
	Nucula sp.	290	21.3	21.3
Four species	Donax vittatus	250	18.3	39.6
>10% each	Tellina fabula	246	18.0	57.6
	Tellina tenuis	153	11.2	68.8
	Cardium edule	70	5.1	73.9
	Venus striatula	67	4.9	78.8
Six species	Spisula elliptica	62	4.5	83.3
>1% each	Ensis siliqua	30	2.2	85.5
	Chlamys varia	22	1.6	87.1
	Thyasira flexuosa	14	1.0	88.1
	Gari fervensis	6		
	Ostrea edulis	5		
	Mactra corallina	4		
	Lucinoma borealis	4		
Ten species	Mytilus edulis	3		
<1% each	Dosinia lupinus	2	2.0	90.1
	Corbula gibba	1		
	Lutraria lutraria	1		
	Mya arenaria	1		
	Venerupis pullastra	1		
	other bivalve specimens	39	2.9	93.0
	Gastropods:			
	Hydrobia sp.	14	1.0	94.0
	Littorina sp.	5	0.4	94.4
	Natica sp.	5	0.4	94.8
	other gastropod specimens	27	2.0	96.8
	Echinoid fragments	33	2.4	99.2
	Bryozoan fragments	1		
	Crustacea:			
	barnacle plates	1	0.6	
	crab fragments	3		
	ostracode valves	3		
	total number of specimens	1,364		99.8

From this somewhat limited sample of the intertidal beach there appears to be a very low diversity of living bivalve fauna (five species) with *Tellina tenuis* the dominant species (85% in the Sutton Spit section and 74·4% in the Wreck section). Abundance is also relatively low with an average bivalve density between 85 per m^2 (Sutton Spit section) and 153 per m^2 (Wreck section), although the total population for the beach area based on these densities appears enormous, estimated at between 288 and 519 million bivalves.

The dead assemblage

This is still bivalve dominated but with a greater diversity and it includes representatives of other invertebrate groups (*Table 7:2*). Most localities yielded at least ten species of bivalves, a gastropod and fragments of echinoids and barnacles. Four bivalve species comprise about 70% of the total assemblage and there is quite a consistent correlation of abundance between species from different localities *(Fig. 7:2.)* However, there are also some interesting differences between the two sections with

Plate 7:1. The foreshore at the north-eastern end of Bull Island with an outgoing tide showing part of the beach morphology from the dunes at the right and the uppermost strand line across to a second strand line and runnel with rippled surface. Note the birds following the retreating tide and feeding at the water line.

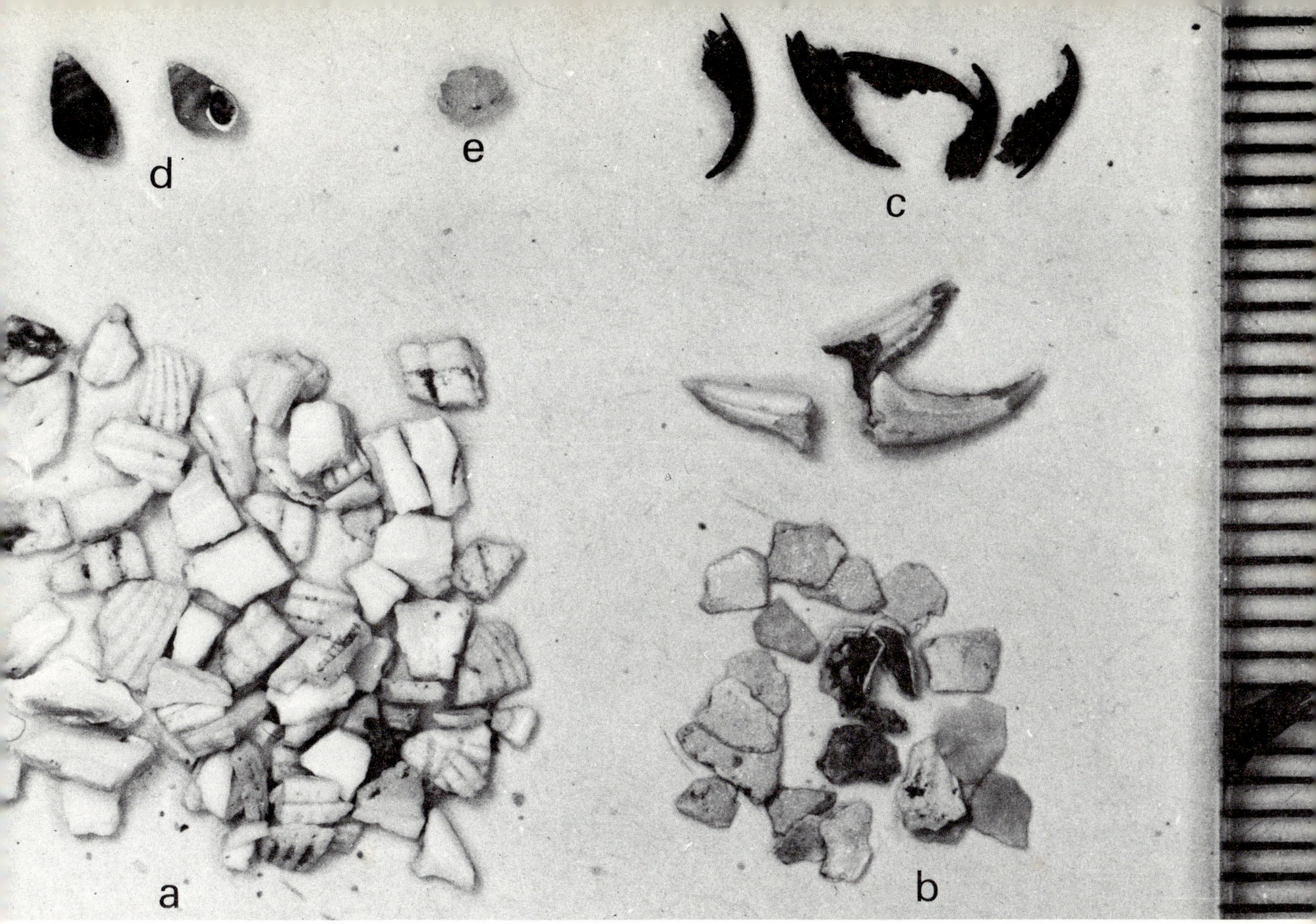

Plate 7:2. A sample of 100 fragments from a seabird cough ball with regurgitated undigested material from its various food sources with hard parts. The majority of fragments are cockles **(a)** accompanied by crab **(b)**, polychaete 'jaws' **(c)**, Hydrobia sp. **(d)** and a sand grain **(e)**. The sample is biased since a single cockle or crab may provide many fragments but a simple polychaete has only two 'jaws'. (Scale in mm.)

assemblages from Sutton Spit including epifaunal species such as *M. edulis; Patella* with littorinids and barnacles. There is a clear indication here of an addition to the beach dead assemblage presumably transported from the nearby rocky shores of Howth.

An obvious consequence of such a faunal mixture is the confusion it could cause to any future palaeontologist trying to reconstruct the original environment of deposition from a study of the contained fossils. Indeed the complexities and contradictions of the situation are amplified considerably if the whole dead assemblage is categorised according to the known life habits and habitats of its contained species (*Table 7:3*).

The bulk of the assemblage is characteristic of the offshore (subtidal) sandy bottom community, which has considerable diversity. The cohesionless substrate allows ease of access to burrowers but also has certain drawbacks as a habitat. The substrate can readily be moved by water currents, eroded in one place and redeposited in another (*Fig. 7:3*). The implications of this for the inhabitants can be severe since they are susceptible to sudden exhumation or burial. If they are not capable of readjusting fairly rapidly and regaining their normal life positions they will not survive. Not surprisingly the majority of the species living in this environment are active burrowers. Several of the bivalves of these sandy substrates are siphonal deposit feeders whilst others are filterers. The deep burrowing echinoderm *Echinocardium cordatum* is also a deposit feeder and the assemblage includes a voracious burrowing gastropod *Natica* sp. which preys upon both bivalves and other gastropods.

The offshore subtidal muddy sand environment is characterised by two small bivalves, *Nucula sp.* and the fragile (thin shelled) *Thyasira flexuosa*. The former feeds on organic rich mud from the sediment water interface. The latter is a deeper burrowing species which has to maintain its respiratory and feeding connections with the water through the sediment without the aid of fleshy siphons and the greater cohesiveness of the muddy sand facilitates this. For quite different reasons both species have a preference for this environment of relatively deep

Plate 7:3. Broken cockle shells collected from the beach surface showing various fracture patterns and the stronger region of the hinge and umbo where the shell is thickened. The fractures tend to follow the concentric growth bands and radial furrows as lines of weakness. Some specimens are still articulated demonstrating the relative strength of the ligament. These also tend to have only one valve broken (the right hand one) at its posterior end through which the siphons extend. This type of breakage is typically produced by mollusc eating seabirds. (Scale in mm.)

Fig. 7:3. Diagram of some of the processes affecting the nature and distribution of the intertidal dead assemblages of shells: current transport with exhumation and burial also predation by birds and gastropods. (Not to scale.)

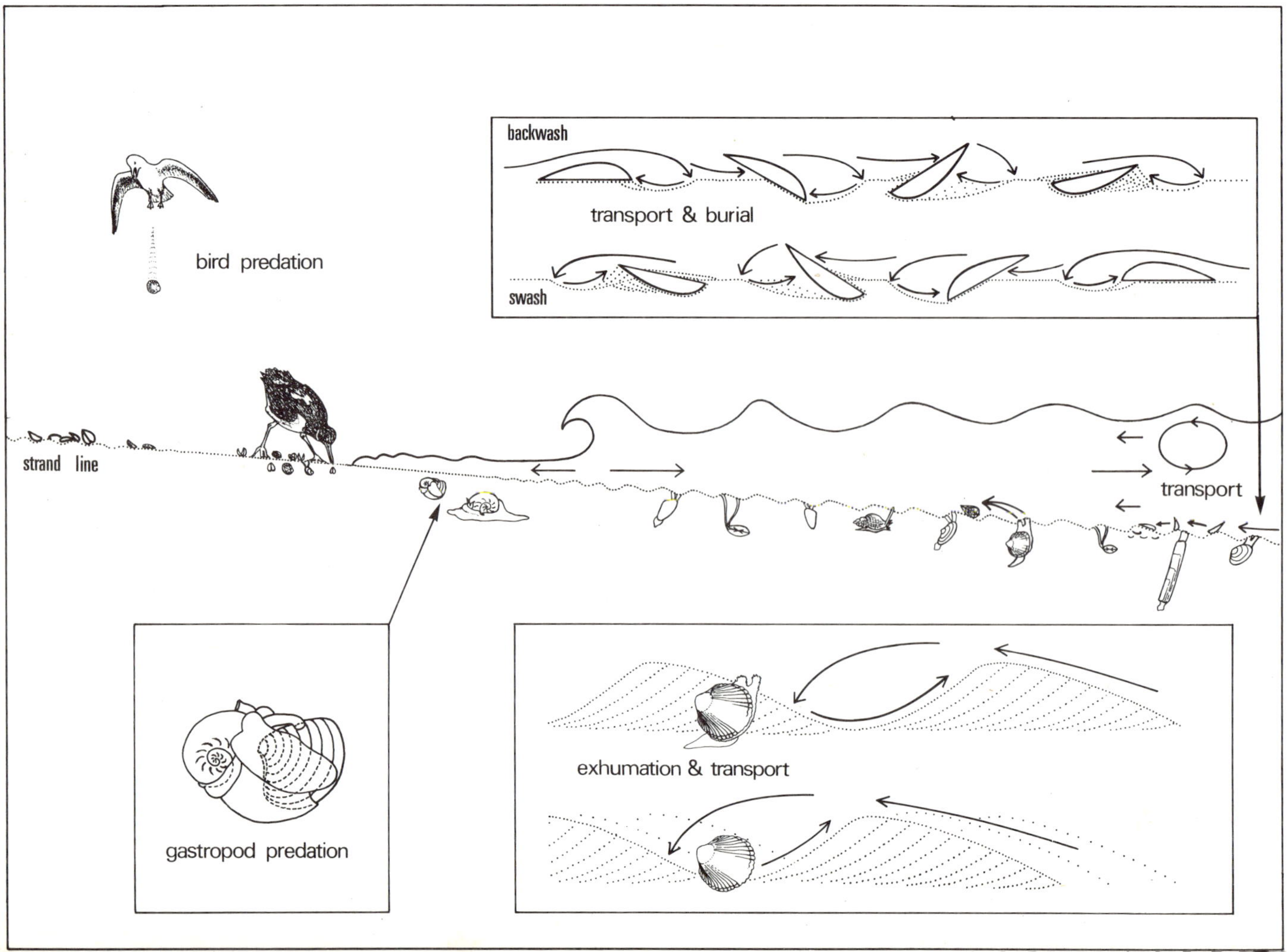

Intertidal

Shallow Rock	Shallow Sand	Shallow Mud
Bivalves	Bivalves	Bivalves
Mytilus edulis	*Cardium edule*	*Cardium edule*
Gastropods	*Donax vittatus*	*Macoma balthica*
Diodora apertura	*Macoma balthica*	*Mya arenaria*
Gibbula cineraria	*Tellina fabula*	Gastropods
Littorina sp.	*Tellina tenuis*	*Hydrobia* sp.
Patella sp.		*Nassarius reticulatus*
Crustacea		
Balanus sp.		

Offshore

Offshore Sand	Offshore Muddy Sand	Offshore Gravel
Bivalves	Bivalves	Bivalves
Chlamys varia	*Corbula gibba*	*Chlamys varia*
Donax vittatus	*Dosinia lupinus*	*Corbula gibba*
Dosinia lupinus	*Gari fervensis*	*Modiolus modiolus*
Ensis siliqua	*Nucula* sp.	*Mytilus edulis*
Gari fervensis	*Ostrea edulis*	*Ostrea edulis*
Mactra corallina	*Spisula elliptica*	*Spisula elliptica*
Tellina fabula	*Thyasira flexuosa*	*Venerupis pullastra*
Tellina tenuis	Echinoid	
Venerupis pullastra	*Echinocardium cordatum*	
Venus striatula		
Gastropod		
Natica catena		
Echinoid		
Echinocardium cordatum		

Table 7: 3. General life habitats of species found in dead assemblages

(below normal wave base) quiet water conditions where such mud can be deposited.

The offshore (subtidal) gravel environment is differentiated from the others only by the presence of larger clasts (pebbles or dead shells) in an otherwise finer grained matrix. The clasts provide a hard substrate that can be colonised by various epifaunal species that could not otherwise survive in this environment (e.g. *Ostrea edulis, Mytilus edulis, Modiolus modiolus, Chlamys varia*). Of these the oyster (*O. edulis*) is cemented itself to a substrate as a juvenile but may outgrow the initial area of attachment and become free living as an adult. Likewise the small clam *C. varia* may lose its early byssal fixation whereas the mussels *M. edulis* and *M. modiolus* retain the anchorage by their byssal threads throughout life.

The intertidal sands have previously been discussed in reference to the sampled living fauna. The only remaining relevant habitat is that of the rocky shore with its predominantly epifaunal biota. Here gastropods dominate bivalves in number of species but not necessarily in overall abundance. Vast populations of *M. edulis* and barnacles frequently colonise rock surfaces. The various species that live here are especially adapted for the particular environmental stresses to be met with, e.g. the form of the limpet *Patella* with its low conical shell and large muscular foot for firm attachment to smooth exposed rock surfaces.

The dead assemblage therefore includes species from a wide range of habitats and represents a mixed or compilation fauna that has been accumulated by physical transportation processes.

Interpretation

If the life habitats and distribution of substrates were not known a quite different approach to understanding the nature of the dead assemblage would have to be attempted. This can be achieved by considering how and why the molluscs die and what happens to them subsequently.

Most of the bivalve species considered here are burrowing forms which live within the sand with a characteristic life position. The shells are generally oriented with their elongate posterior ends uppermost, the long axis and commissure of bivalves being at a high angle to the sea bed. This is quite different from the mechancial angle of rest of an empty shell on the sea bed. Consequently it is not difficult to distinguish between life and death positions of most bivalves.

Predation

Exactly how and why bivalves die is not often known but some indication can be gained from an examination of empty shells. In the intertidal environment bivalves are particularly susceptible to bird predation. Oyster catchers are well known as very effective and voracious consumers of bivalves especially cockles (*Fig. 7:3*). A single bird is capable of devouring over 200 cockles a day and it is evident

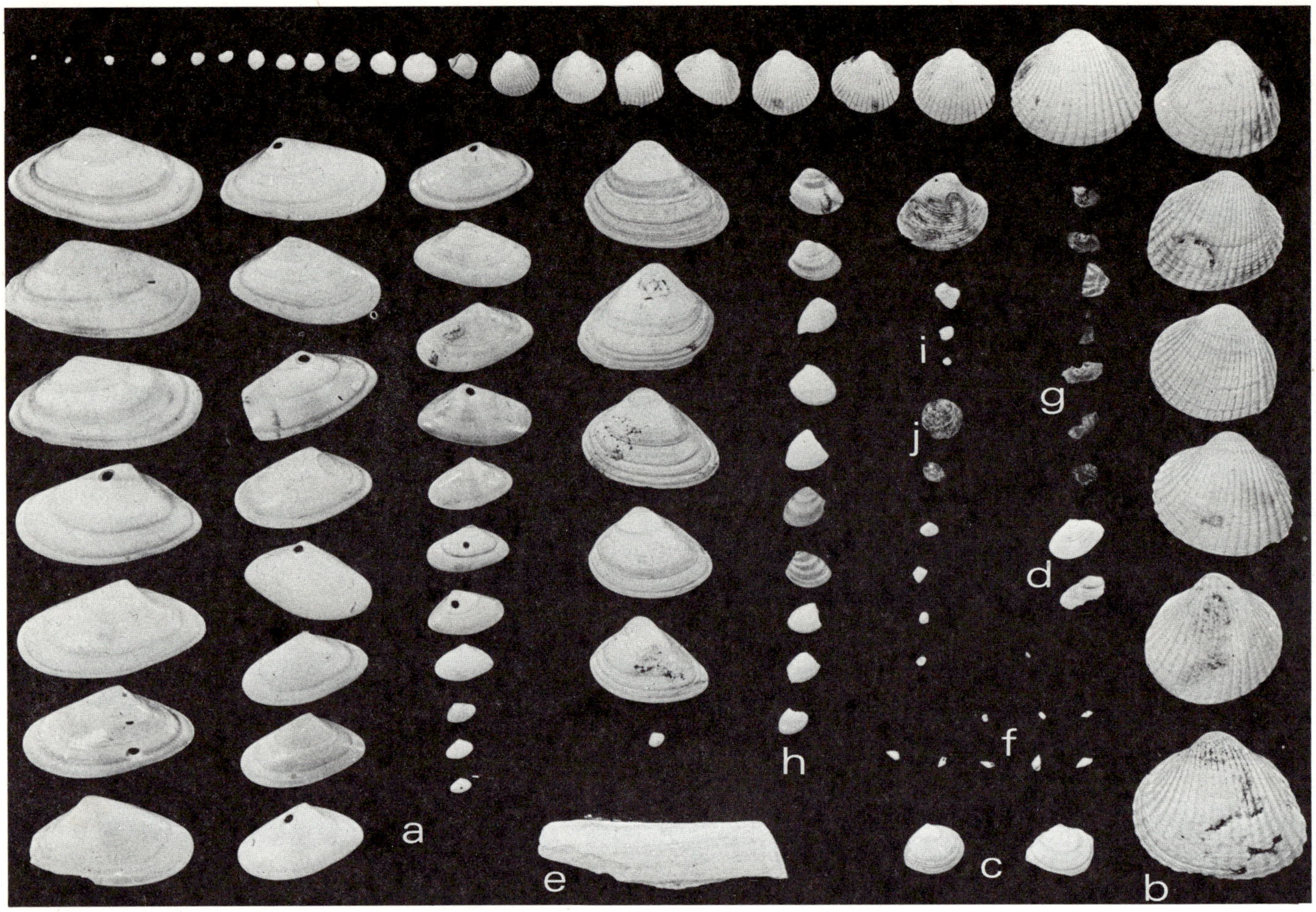

Plate 7:4. A sample count of 100 'particles' (i.e. whole or nearly whole identifiable shells) from a coarse fraction (greater than 2 mm particle diameter) within the beach sand. Represents a low diversity dead assemblage of predominantly intertidal to low subtidal bivalves with some additional shelly invertebrates from other habitats. The **(a)** *Donax vittatus,* **(b)** *Cardium edule,* **(c)** *Macoma balthica,* **(d)** *Tellina sp.,* **(e)** *Ensis sp.* and **(f)** *Hydrobia sp.* live within the immediate vicinity whilst the remainder come from diverse other habitats ranging from rocky foreshore **(g)** barnacles to offshore muddy sand **(h)** *Nucula sp.* As can be seen from the number of bored *D. vittatus* shells, this species suffers greatly from gastropod predation. Other identifiable species are **(i)** *Venus striatula,* **(j)** *Ostrea sp.*

that cockle populations can be seriously depleted by winter feeding flocks. The birds' method of killing and eating the bivalves leaves quite distinctive traces on the shell. As a shallow burrower, a cockle with its short fleshy siphons is just visible at the sediment surface to the keen eye of an oyster catcher. The bird drives its long sharp beak swiftly and accurately into the siphons and kills the cockle, probably severing one or both of its adductor muscles at the same time. The valves can then be easily prised open by a twist of the beak and the flesh eaten. The action of the beak frequently chips the posterior siphonal margin of the shell and sharp angular fractures are produced in one of the valves by the twisting motion. These features can commonly be observed in freshly killed cockles (*Plate 7:3*).

Other bird species, particularly the various gulls and hooded crows, feed on exhumed bivalves washed up on to the beach in strand line deposits.

Bivalves are no safer from predation in the subtidal environment since there they are vulnerable to attack by fish, crustacea, starfish and gastropods as well as various parasites. Most of these predators do not leave diagnostic traces on the shells but gastropods do.

Two species are particularly successful and common burrowing predators on bivalve infaunas and actively seek out their prey within the sand. The gastropod envelops its hapless prey with its extensive muscular foot and whilst thus firmly attached slowly drills a neat hole through the shell using its horny rasping radula and enzymes (*Fig. 7:3*). With the hole complete the extensible proboscis is protruded through into the shell to devour the flesh. Such predation clearly leaves its mark and an examination of the dead assemblage shows that three bivalve species are particularly prone to this type of attack (*Plates 7:4, 13 a, c. Table 7:4*).

	Sutton Spit Section		Wreck Section	
	borings / sample size	% bored	borings / sample size	% bored
V. pullastra	1/4	25	0/1	0
S. elliptica	20/106	18.9	10/62	16.1
V. striatula	2/11	18.2	13/67	19.4
D. vittatus	9/166	5.4	13/250	5.2
Ostrea sp.	4/120	3.3	0/5	0
Nucula sp.	5/230	2.2	18/290	6.2
M. edulis	1/47	2.1	0/3	0
T. fabula	5/359	1.4	2/246	0.8
C. edule	3/323	0.6	0/70	0

Table 7:4. Counts of holes in bivalve shells bored by predaceous gastropods. It appears that some bivalve species are more vulnerable than others.

Plate 7:5. Current oriented shells on the foreshore swash zone showing chevron marks produced by current scour and picked out by the differential sorting of small white shell fragments from the sand (current movement from top to bottom). The slightly different density of the calcareous shell material from the largely quartzitic sand results in slightly larger shell fragments being transported with a particular sand grain size at any given water velocity. Also shown is the characteristic shell fracture pattern resulting from human activity.

Stranding

Apart from predation it is likely that the most common cause of bivalve mortality is exhumation with rapid transport beachwards and subsequent stranding leading to death by exposure or predation (*Fig. 7:3*). The mechanical processes of transport, degradation and burial of exhumed shells are complex and can only be briefly referred to here. Much depends on shell shape and size, the surface area to weight ratio and the initial orientation of the shell on the sea bed. Articulated globose shells and dis-articulated concave upwards ones are likely to be transported by water currents, whereas those oriented concave downwards are more likely to be buried by the same current. However, since the concave upwards position is an unstable one for a single shell it is quite likely to be tipped over during transport and subsequently come to rest in the more stable concave downwards position (*Plates 7:5, 6. Fig. 7:3*). Then it is more likely to be buried.

Time lapse since death

There are also various criteria by which some judgment can be made of the relative time lapse since death (*Table 7:5*). Animal tissue lasts at most only a few weeks before decaying or being removed by scavengers. The internal nacreous lustre of many bivalve shells is gradually dulled and lost over two or three years subsequent to death.

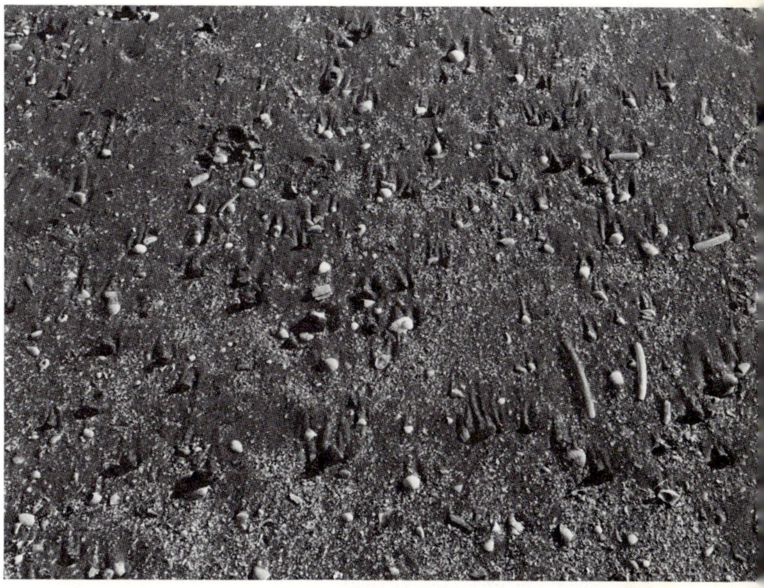

Plate 7:6. The swash zone of the upper beach with a considerable number of disarticulated shells, shell fragments and pebbles forming the coarsest fraction of the sediment, which has been accumulated by strong wave formed currents. There is a distinct orientation of the shells (especially noticeable in the elongate *Ensis* sp.) and scouring down-current from the larger objects shows that the current (probably a strong backwash in this instance) moved from bottom to top of the photograph, i.e., seawards, down the beach. Notice also that the majority of the shells are in the stable position, i.e., concave downwards.

North Bull Island – Dublin Bay

	DAYS	WEEKS	MONTHS	YEAR	YEARS
Tissue	Tissue decays rapidly with microbiological activity or is quickly removed by scavengers unless desiccated	Desiccated muscle fibres adhering to adductor muscle scars may last some weeks or possibly months			Very rarely some soft parts (e.g. gut) can be preserved by pyritisation when the animal is entombed in anaerobic conditions, e.g. black foetid mud
Periostracum	Only well developed in some species and will degrade fairly rapidly when dried, especially through mechanical abrasion			Where thick, e.g. *Modiolus* may last a year or so	
Nacre	The internal pearly lustre becomes gradually dulled by mechanical abrasion or chemical solution with acid ground water over a period of a year or so, but may last slightly longer on the muscle scars				
Colour	Gradually fades over a year or so especially with exposure to strong sunlight, rate depends on initial intensity of coloration				In special conditions strong coloration can be fossilised and is known in fossils from as far back as the Devonian
Ligament	This organic material is similar in composition to periostracum but has a slightly greater preservability				In special conditions it can be fossilised and the oldest known is of Permian age
Disarticulation	Depends on the size of the ligament attachment area relative to shell weight and strength, also the energy conditions under which the animal died and was buried. In a high energy situation, e.g. exposed sandy or pebbly beach shells are rapidly disarticulated				In a low energy situation shells may be buried in the life position with valves still articulated
Fracture and Abrasion	Fracture and reduction in shell weight by mechanical abrasion depends on the energy conditions of the situation in which the animal died and was buried; also, the strength and thickness of the shell with its size:weight ratio. Fracture and rapid abrasion will occur in high energy conditions where there are available 'tools' and abrasive material, i.e. rocks and sand. Experiments have shown that *Mytilus* valves show a 15% weight loss after 1 week continuous tumbling in wet fine sand and 60–80% after 10 weeks which is equivalent to being transported some 2,000 km. In natural situations net movement over such a distance is unlikely and would take much longer. *Mytilus* 15% weight loss after 1 week continuous abrasion — experimental — 60–80% loss after 10 weeks N.B. Initial fracture is frequently produced by predation especially by mollusc eating birds.				
Epifaunal Encrustation	Maybe either constructive and/or destructive in that epibionts may protect the shell from mechanical abrasion but also may actively bore into the shell, thus weakening it. Colonisation may be and frequently is initiated during the life of the host and it is not always possible to differentiate pre- and post-mortem encrustation except for instance where there has been an invasion of the living space. Growth rates of the epibionts are not generally known but for calcareous secreting organisms are unlikely to exceed 3 mm per 100 days.				

Table 7:5. An attempt to assess the relative rates of various post-mortem processes which effect visible changes in shells and thus can be used to make some estimate of time-lapse since death of their inhabitants.

Plate 7:7. Upper foreshore with strand line deposit in swash zone after a winter storm. Scattered amongst the clumps of seaweed are numerous large still articulated valves of the prickly cockle *(Cardium echinatum* **(a)**), with a few other bivalves, gastropods (e.g., *Buccinum undatum* **(b)**, and *Tricla lignaria* **(c)**) flotsam and jetsam. The flesh has mostly been scavenged from the bivalves except for traces of muscle still adhering to the areas of attachment on the inside of the shell. These bivalves have only recently been exhumed from their shallow burrows in the sand by erosion of the substrate during a storm, transported some distance inshore from their habitat, stranded on sand too firm to burrow in and there killed by predators (probably birds). The shells are still held together by the strong elastic ligament. Although large and heavy such shells can be moved quite easily by water currents because of their 'unstreamlined' and large surface area.

Plate 7:8. A storm assemblage of large shells from different habitats assembled as a strand line deposit. The shells show a number of post-mortem features indicating a considerable variation in the length of time since death. Amongst the oldest are the oysters **(i)** and *Buccinum* shells **(j)** with their well developed epifaunal encrustrations. Where these epibionts have invaded the living space of the host colonisation has clearly been post-mortem. However, since these species are sediment surface dwellers some of the external encrustation may well have developed during the host's life without harming it. Several of the bivalve shells are still articulated but the only remaining flesh is restricted to traces of muscle fibres adhering to the areas of attachment on the inside of the *Lutraria lutraria* shells **(a)** (a specimen of which has been disarticulated to show this). The latter also have their thin periostracum partially preserved indicating that they are not long dead (a week or so). The articulated specimens of *Artica islandica* **(b)** and *Mytilus edulis* **(c)** have their thick periostraca almost completely intact but retain no traces of flesh and thus have been longer dead (several weeks). Another large single valve of *A. islandica* is much older having completely lost its periostracum and ligament (several months or more) **(d)**. The gastropods include three specimens of *Tricla lignaria* **(e)** a burrowing predator from relatively deep water off-shore. The periostracum has been almost completely removed from their thin fragile shells. Thus it would seem that whilst they may have been dead for some time it must have been in undisturbed water for they are quickly broken up on the beach. This similarly applies to the echinoids. There is a single specimen of a regular form **(f)**, from a rocky foreshore habitat with most of its spines removed but some traces of tissue, also two and a half specimens of the irregular burrowing echinoid **(g)** *Echinocardium cordatum* from which all spines and traces of flesh have gone. Of the crustacea there is the separated carapace and body of a swimming crab **(h)** (Portunus depurator) with some flesh and articulated limbs intact indicating recent death and probably bird predation on the beach. Other species illustrated are: **(k)** *Venerupis pullastra,* **(l)** *Nassarius sp,* **(m)** *Cardium echinatum,* **(n)** *Mactra corallina.*

The degree of preservation of the periostracum (the outer organic layer on the shell) which is well developed in some bivalve species (*Plates 7:8, 14, 15*) provides a good guide to the relative amount of abrasion a shell has suffered through transport.

It is for instance a reflection of considerable transport and postmortem time lapse that none of the abundant *Nucula* shells of the dead assemblage ever have soft tissue preserved. Also, it is only after the most violent of winter storms that specimens are found with the periostracum intact and the valves still articulated by the ligament. Preservation of shell pigmentation depends on various factors, including the intensity of the original colour, its hue and bleaching by subaerial exposure. Some species (e.g. *Chlamys*) may retain their colour for a year or so.

A quite different aspect of postmortem change is seen in the degree of colonisation of a shell by epibionts (*Plate 7:9*). This can be both a constructive and a destructive process in that such a covering may slow down shell abrasion but at the same time some of the epibionts may be actively removing shell material by boring and/or chemical solution. These epibionts are only useful in this taphonomic context when they invade the living space of the bivalve and are therefore clearly colonising it after death. The growth rates of the common epibionts are as yet unknown or so variable as to be almost useless for accurate dating, though it is likely for example that it will take several months for the establishment of organisms such as barnacles, bryozoa and cementing bivalves. In the intertidal zone barnacles grow at a rate of 1 – 2 mm per 100 days.

The rate of mechanical destruction of shells is complicated by various factors such as the presence of epibionts, shell thickness and architecture, surface area: weight, biological degradation, sediment particle size and sorting.

In the sedimentary environment typified by the littoral sand flats of Bull Island with their well sorted fine grained sands (*Plates 7:10, 11*) some generalisations about abrasion can be made. Experiments have shown that *M. edulis* valves suffer a 15% weight loss by abrasion after 185 hours' continuous rolling in wet fine sand. After some 1,800 hours there has been a 60 – 80% weight loss and this is equivalent to continuous transport over some 2,000 kms for nearly three months.

Plate 7:9. Three large molluscs from the strand line with well developed epifaunal encrustations: **(a)** The *Buccinum* has clearly been dead for a year or so allowing colonisation of the living chamber by the calcareous tube worm (*Pomatoceras* sp.) which is also found on the shell exterior associated with barnacles. These latter may have initially colonised the shell whilst the gastropod was still in occupation. The shell has also been bored (as can be seen in the columella and apertural lip) by the ectoproct bryozoans *Terebripora* sp. and *Polydora* sp. **(b)** The still articulated shells of *Venerupis pullastra* have a barnacle epifauna restricted to the exterior surfaces. This may have developed during the life of the bivalve since it lives on the sediment surface and may be byssally fixed. **(c)** This internal surface of a large oyster has a massive encrustation of calcareous tube worms with some bryozoans. The oyster has been dead for more than a year with the encrustation adding considerably to the weight and thickness of the shell thus making it more resistant to mechanical abrasion.

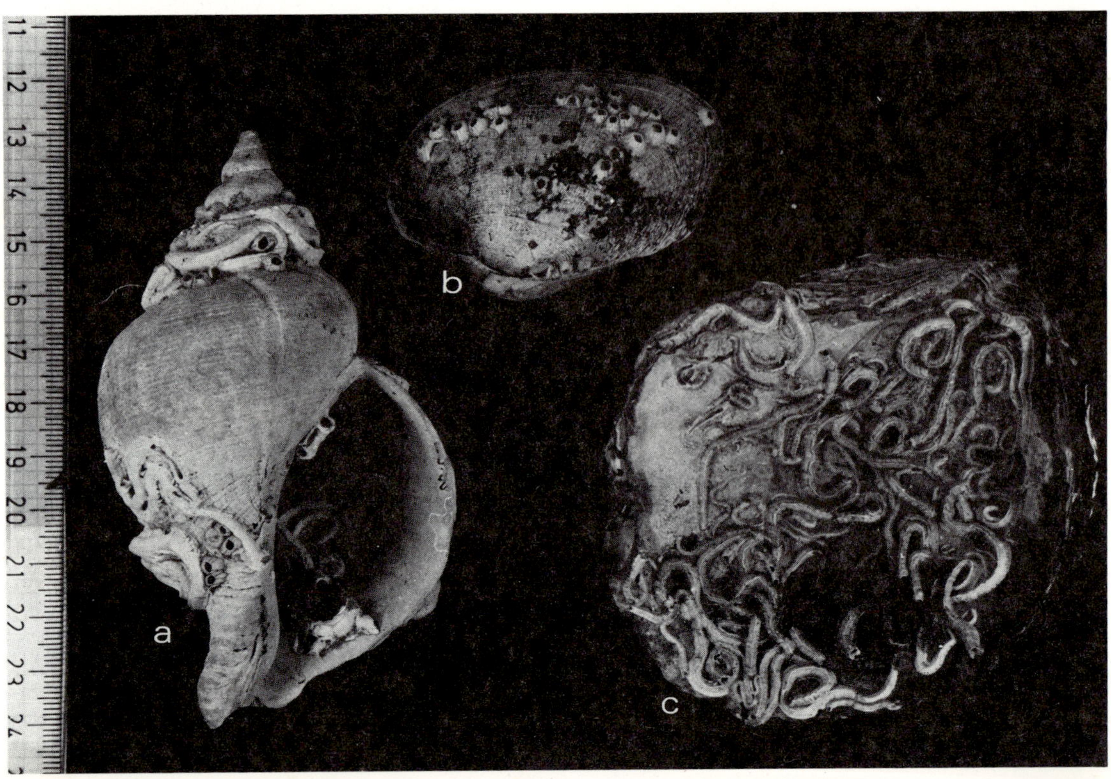

Plate 7:10. Sample of sand from the foreshore separated by sieving into seven size classes **(a)** > 2 mm, **(b)** 1 mm, **(c)** 0·5 mm, **(d)** 0·25 mm, **(e)** 0·125 mm, **(f)** 0·062 mm, **(g)** < 0·062 mm. The bulk is of fine sand between 0·25 – 0·125 mm grain size diameter. The coarser fraction is largely made up of shells and shell fragments, whilst the fine 'tail' is composed of silt. This demonstrates the bimodal origin of the deposit. The predominantly inorganic quartz sand is well sorted and derived from the reworking of Pleistocene deposits in Dublin Bay and the Irish Sea. This is joined by the locally produced calcium carbonate fraction from the 'shelled' invertebrates.

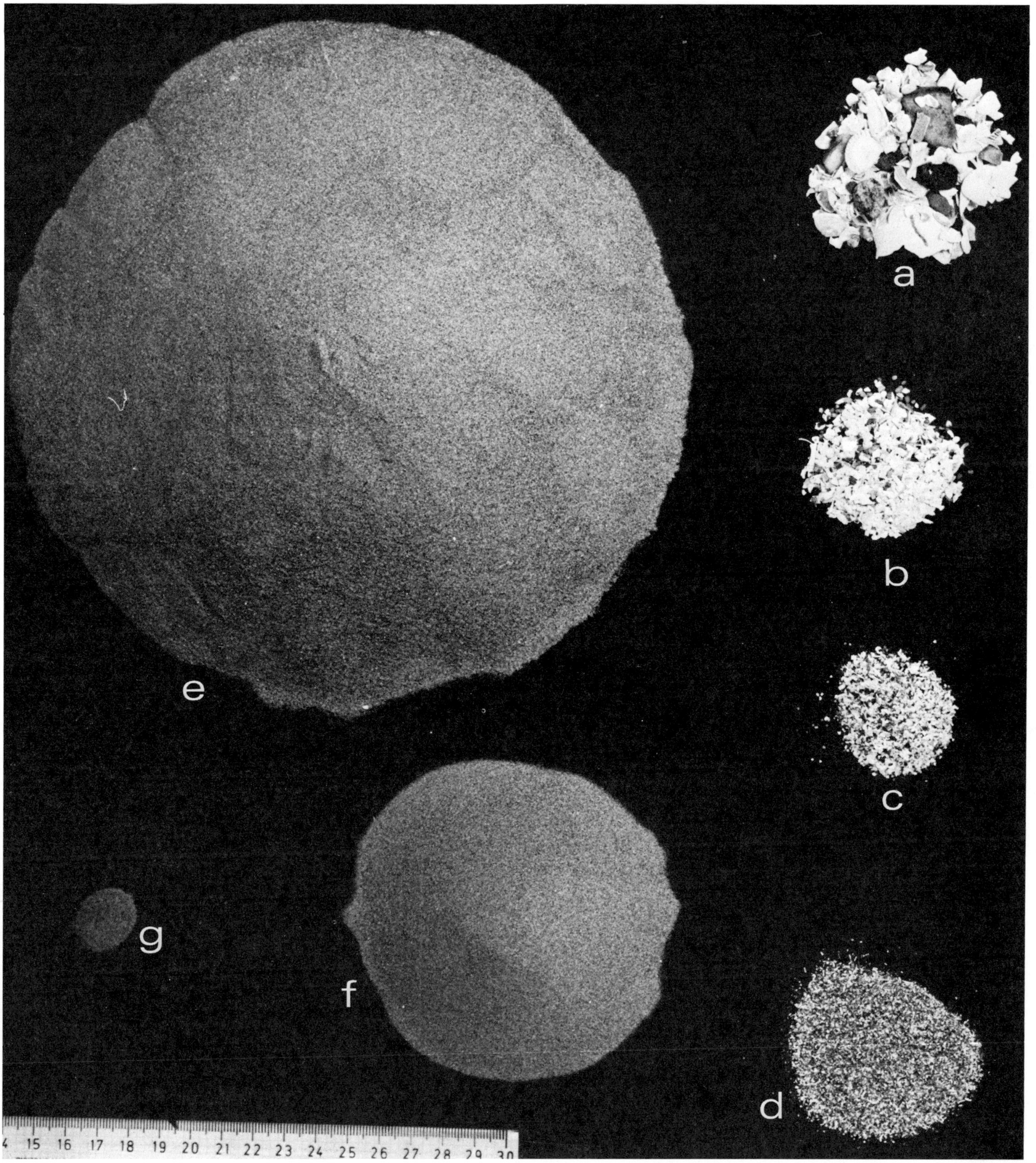

Plate 7:11. Comparison of the carbonate fraction separated into four size classes by sieving: **(a)** > 2 mm, **(b)** 1 mm, **(c)** 0·5 mm, **(d)** o·25 mm. The largest grade includes a considerable variety of shells from different bivalve species and their fragments along with some large pebbles. These represent grains that can only be moved by relatively strong currents and consequently tend to be transported within the breaker zone where they are also fragmented. The next finer grades (2 and 3) include shell fragments and small shells, both juveniles of the species found in the larger fraction and adults of smaller species. Gastropods frequently occur in this size class because a coiled shell of this size is very resistant to fracture and abrasion. The smallest of these carbonate dominated fractions includes various 'shells' not found in the other fractions, e.g., forams, ostracodes and spines from *Echinocardium cordatum* because this is their maximum growth size. (Scale in mm).

Articulation of bivalves depends on a resilient organic ligament. Its elastic strength persists for a few months after death and commonly forces the valves to open once the adductor muscles relax or are lost (*Plate 7:7*). However, continued articulation of the valves also depends on the ligament strength and especially its effective area of attachment relative to shell weight and size. Experiments have shown that 50% of a sample of 100 or so specimens of *M. edulis* were disarticulated after some 20 – 30 hours' continuous rolling and that after 100 hours all but 10% were separate and finally all were disarticulated after 180 hours.

The Bull Island assemblage includes thin shelled species (e.g. *T. flexuosa* and the tellinids) which would appear to be relatively fragile. However, because they are also small and light they are more resistant to fracture and abrasion than expected. An illustration of this effect can also be seen in the degradation of the echinoid *E. cordatum* (*Plate 7:7, 8a*). This has a very thin fragile and somewhat globose calcium carbonate test (up to 5 cm in diameter) and numerous fine spines (less than 1 mm in diameter and up to 10 mm long). Only occasional fragments of the test are found whereas the spines are quite common. In places both on the sand surface and within it there are bedding planes covered with countless spines that have been differentially sorted by water currents (*Plate 7:12*). The spine shape, size and fine meshwork structure (*Plate 7:13d*) makes them apparently too light to be readily broken down.

As a result of all these different factors it is very difficult to age a shell on the basis of the amount of abrasion it has received or to assess the total bias to the record of the relative destructibility of different species. The shells of Bull Island beach become mechanically degraded by fracture but show few signs of abrasion (compare *Plates 7:2, 6, 8, 11*). This may be due to either the sheltered conditions of the bay, the lack of abrasiveness of the fine grained sand, rapid burial or a combination of these factors.

Some assessment of shell burial relative to transport can be made. The dead shells on the beach surface has a density distribution of approximately 30 shells per m^2 (equivalent to 15 live individuals). Compared with this the density of the living population is much higher at 85 – 153 per m^2. Since the dead epifauna includes shells recruited both from the adjacent sands and the other habitats of the bay the low density would seem to indicate rapid burial of empty shells, at least within the intertidal zone. This contention is supported by the lack of abrasion, epifauna or cumulative shell ridge at high water mark. There is however some evidence for a general shoreward movement of shells which become dispersed throughout the intertidal zone.

Another aspect that can only briefly be discussed here is the role of strand line deposits. Winter storms are particularly effective in exhuming an exotic fauna from the sea bed and transporting it onto the beach where it is left stranded amongst a variety of flotsam and jetsam (*Plates 7:7, 14*). Commonly the assemblage consists of recently dead specimens of large bivalve species accompanied by a few gastropods and echinoderms. Although the shells of the bivalves are generally still articulated only the mytilids retain their soft parts.

It is difficult to assess the importance of these strandline accumulations within the overall dead assemblage but few of their member species are encountered with any frequency. This suggests that they tend to be dispersed and overwhelmed by the much greater abundance of the common species of the dead assemblage. It is also perhaps significant that the shells of the dead assemblage are generally less than 2 cm long whereas those of the strandline deposits are 3 – 12 cm long. This probably partly reflects the much greater transporting powers of storm waves.

Finally it appears that the potential fossil record of life in the intertidal sands of Bull Island will be heavily biased towards a minority member (the Bivalvia) of the overall biota. Furthermore, even within this group the record will be distorted by a mixture of species from a wide range of habitats found over a considerable area and assembled in the beach sands.

However, if a palaeontologist studying the assemblage in the future were to assess the condition of preservation of the individual shells and their distribution it should be possible to realise the transported nature of the cumulative assemblage. With some knowledge of the general habits and habitats of different members obtained from a study of shell form it should also be possible to separate to some extent the different components of the assemblage.

Plate 7:13. Scanning electron microscope photographs of various organic components from the 0·25 – 0·5 mm grain size fraction. All traces of tissue have been removed by predators or scavengers and decay. Thus these small 'shells' join the inorganic sand as calcium carbonate (except for the calcium phosphate fish vertebra) clastic components.

(a) A bivalve shell with a bevelled hole neatly drilled by a predaceous gastropod.

(b) *Ostrea* sp. showing marked change in growth form produced by the transition from a free swimming to an attached mode of life with spat settlement.

(c) An ostracode, a 'bivalved' crustacean with its two valves still articulated and bored near the hinge by a predaceous gastropod.

(d) One of the three 'blades' from an echinoid pedicellaria (small grasping and snapping structures between the spine bases, which keep the test clean and free of detritus). Note the meshwork structure of this calcium carbonate structure, so characteristic of the echinoderms.

(e) The two valves of an ostracode opened to show the hinge line and dentation for articulation.

(f) A mixed assemblage of foraminifers (protozoa with coiled and chambered shells), a fragment of a 'stick' bryozoan colony and a fish vertebra.

(g) A fish vertebra with all but one of the arches broken off leaving the tubular centrum as the part most resistant to abrasion.

(h) Juvenile bivalve shells, *Cardium* sp. in the foreground and *Nucula* sp. in the background, the former showing characteristic fracture patterns along the radial ribs and furrows and concentric growth lines produced by purely mechanical stress.

Plate 7:12. Sediment surface in the swash zone with a mixture of shell fragments, small molluscs, ostracodes, foraminifers, echinoderm spines and inorganic sand grains (largely quartz) also three large bivalve shells. The elongate grains, especially the echinoderm spines have been aligned subparallel to one another by the depositing current.

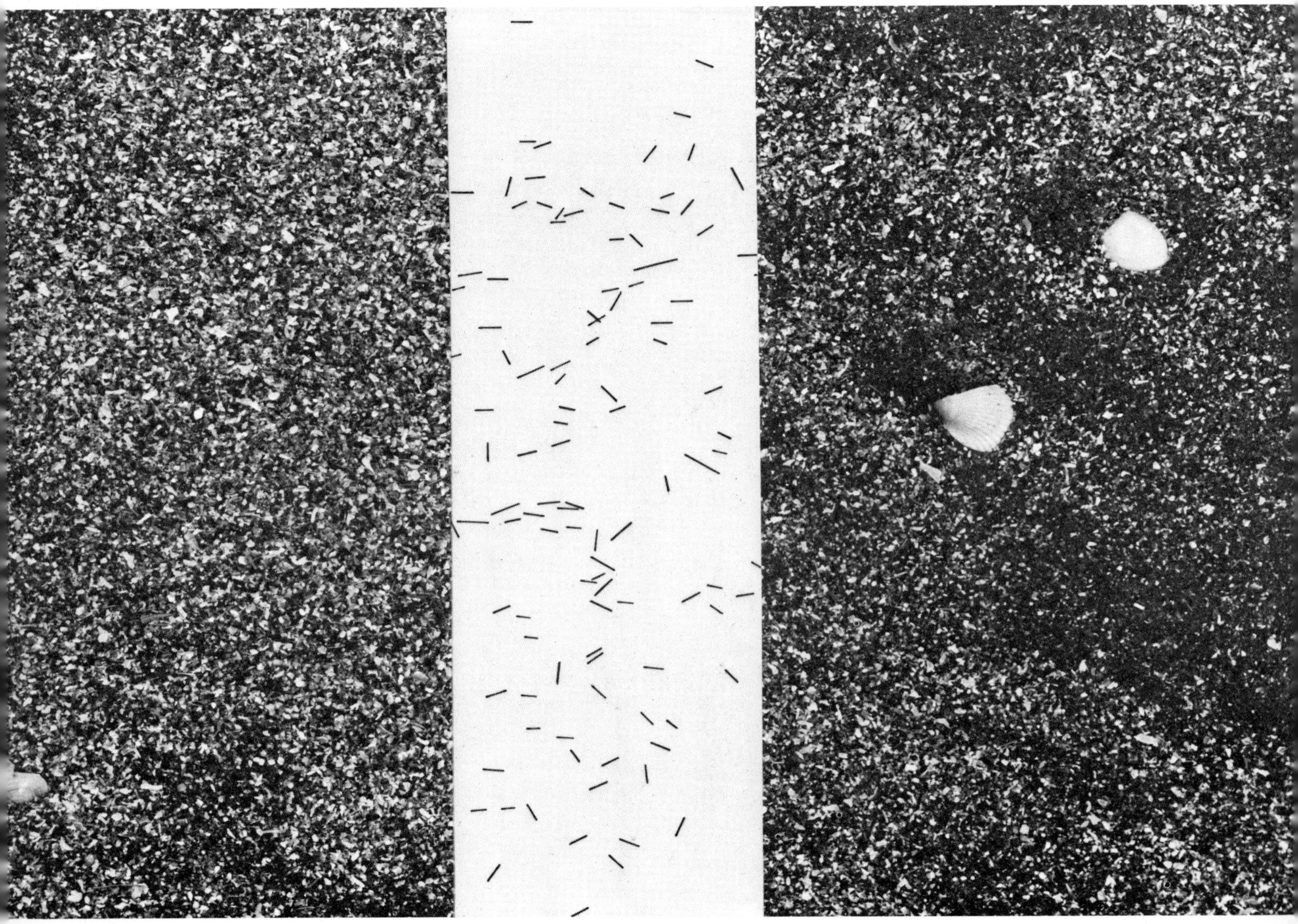

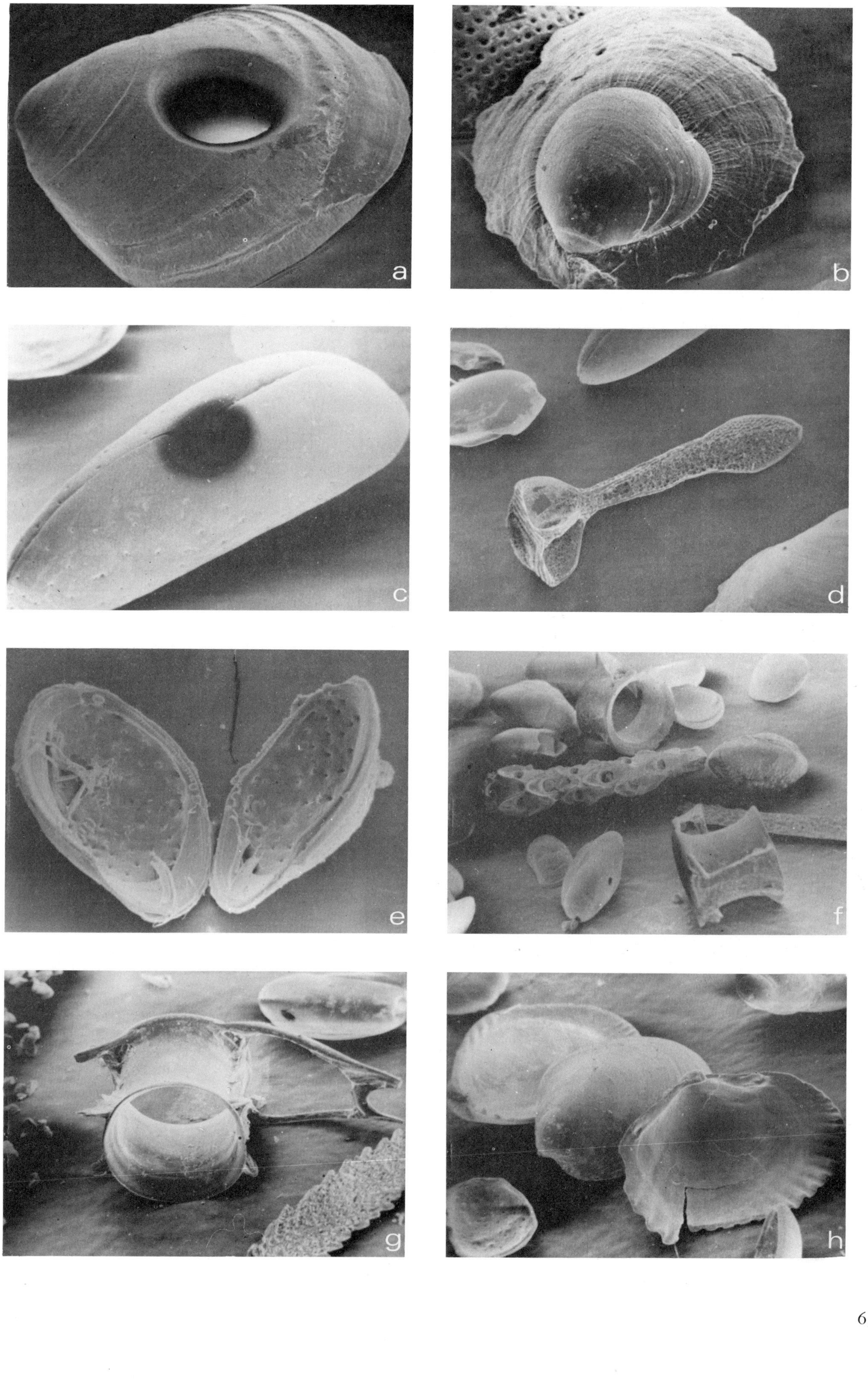
a
b
c
d
e
f
g
h

Plate 7:14. A mixed assemblage of burrowing and epifaunal animals gathered together from 100 sq. m. of a strand line deposit. This includes (a) live large mussels (with thick black periostracum), (b) live cockles, (c) recently dead starfish, (d) the bivalve *Gari fervensis,* (e) empty *Buccinum* shells with barnacle epifauna, (f) razor shells *(Ensis siliqua),* (g) disarticulated *Lutraria lutraria* and (h) *Donax vittatus* also a (i) dogfish egg case.

Plate 7:15. A sample of 100 specimens from the strand line with many of the bivalves still articulated and members of the local infauna, e.g., **(a)** *Cardium edule,* (N.B. bird predation damage to posterior end of the shells, **(b)** *Tellina* sp., **(c)** *Spisula elliptica* T. fabula and **(d)** *Ensis siliqua.* The latter species demonstrates the relative strength of the ligament since the large brittle shell fragments readily but the tough ligament will hold at least the adjacent parts of the shell together. The gastropods have, however, been assembled from other habitats e.g., rocky foreshores **(e)**, *Littorina littorea,* offshore sand burrowers **(f)**, *Natica sp.,* **(g)** *Aporrhais pespelecani,* **(h)** *Tricla lignaria.* Many of the gastropods show considerable abrasion and breakage indicating that they have been dead for some considerable time. The assemblage includes several pebbles most of which are sinter which was probably dumped in the Bay years ago by coal burning ships. Also, there are exhumed tubes of the sand mason **(i)** *Laniche conchilega.* Other identifiable shells are: **(j)** Neucella sp., **(k)** Nassarius sp., **(l)** Mactra corallina.

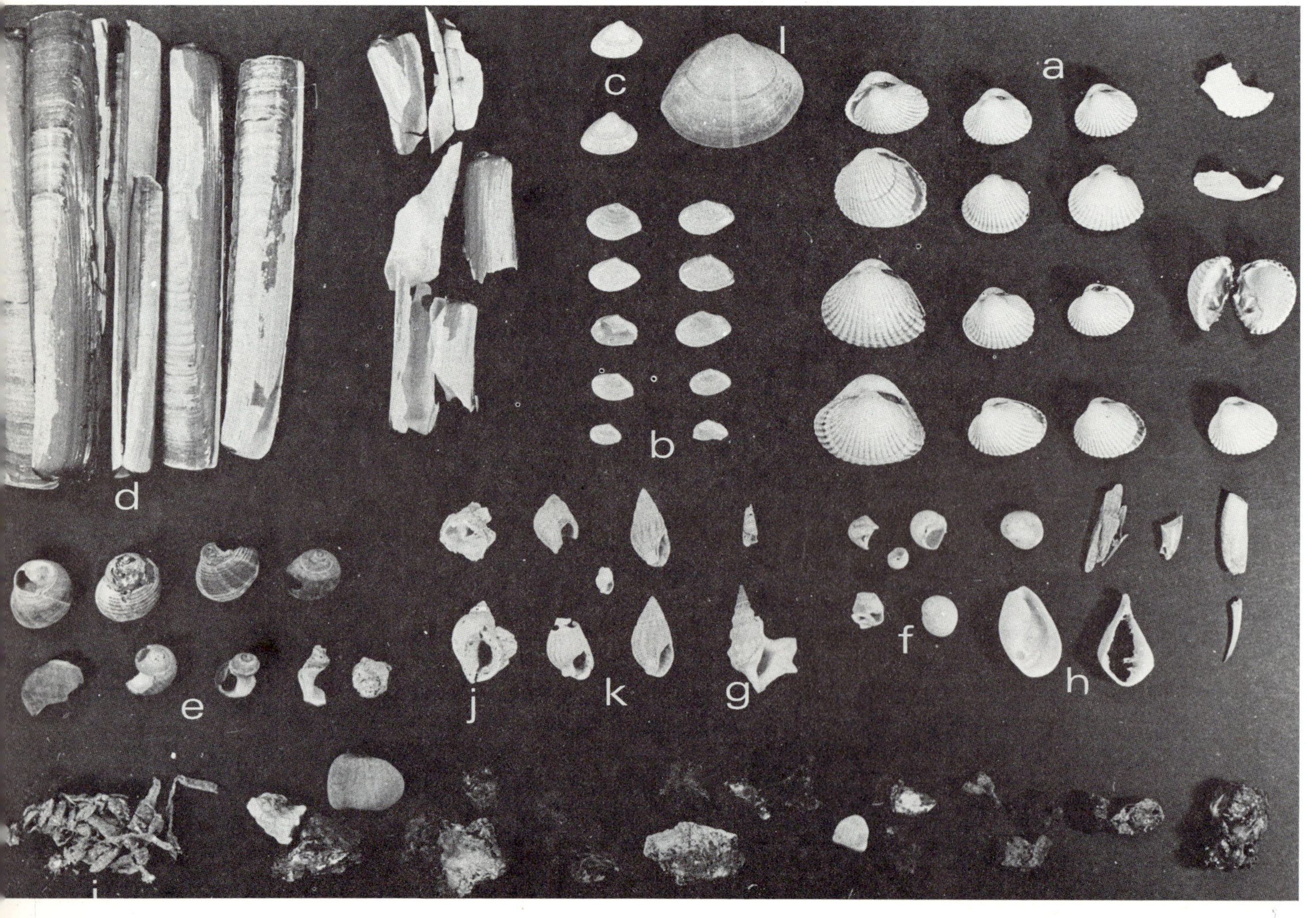

Chapter 8
THE BIRDS OF BULL ISLAND

C. D. Hutchinson *Irish Wildbird Conservancy*

The Bull Island has probably received more attention from bird-watchers to date than any other site in the country. From the records of those who were active in the nineteenth century it is clear that the attraction of the island for waterfowl was recognized early on by the keen ornithologists. However the particular affection of many people for the area stems from the work of Fr. P. G. Kennedy, S.J. He realised Dublin's good fortune in having an area so rich in bird life within a short journey from the city centre and devoted much of his leisure time to establishing a bird sanctuary there, to preserving it from encroachment and to publicising its birds. Since his book was published in 1953 there has been an explosion of popular interest in ornithology all over Ireland and many have come to know the Bull Island and its birds for the first time.

Even now, however, knowledge of the bird life is by no means complete. A start has been made by the completion of surveys on the numbers of wildfowl and waders that occur and by the accumulation of more casual records in the annual bird reports. But one feels that the surface has hardly been scratched and the summary of the status of the birds that follows is based entirely on the work of amateur bird-watchers. Kennedy's observation that 'here at Dollymount . . . would be opportunities for students to tackle some of the many bird problems that remain unsolved' is still valid, especially for the sciences of bird behaviour and ecology.

Common wildfowl species

The most striking feature of the North Bull is the density of wildfowl to be seen. Small wading birds are quite familiar to anyone visiting the coast but large flocks of ducks and geese are not normally seen at such close range. From late November to early February, when the tide is out, the mud is darkened by great packs of wildfowl which are generally found in certain specific areas, depending on their feeding preferences. The brent geese at this time of the year frequent the Sutton and Kilbarrack end of the channel where the *Zostera* grows *(Plate VII)*; shelduck are most scattered but the majority feed along the channel from Raheny to Kilbarrack: wigeon *(Plate 8:1)* congregate most densely off Raheny about 200 m. north of the causeway. Here too the teal pack together amongst the *Salicornia* in flocks of up to 500 birds, though they also scatter in small parties along the channel. Like the wigeon they move regularly away from the North Bull to feed at Clontarf. Pintail tend to flock between Raheny and the causeway and most fly over the causeway when the tide is in, to the saltmarsh on the Dollymount side. Shoveler and mallard are the most particular of all. Shoveler are found just north of the causeway and mallard occur in a small party a few hundred metres further on, near the kink in the channel where a small freshwater stream enters.

Just as the distribution of the wildfowl species on the North Bull channel is quite distinctive, the seasonal pattern of their occurrence is similar each year. There may be temporary differences for the birds move in small numbers to and from the nearby estuaries of Rogerstown, Malahide and Baldoyle. The pattern that emerges is related to each species' life-cycle. Many of the wildfowl breed in northern Europe and Greenland *(Fig. 8:1)* and come south in autumn as food and climate begin to deteriorate. Our wigeon nest in Iceland, Scandinavia and Russia, while teal, shoveler and pintail breed across much of north-west Europe. The brent geese are found in Greenland and Artic Canada and many pass through Iceland on their migration. Shelduck are the exception in that they nest in Ireland although they do leave temporarily to moult in huge flocks in Heligoland.

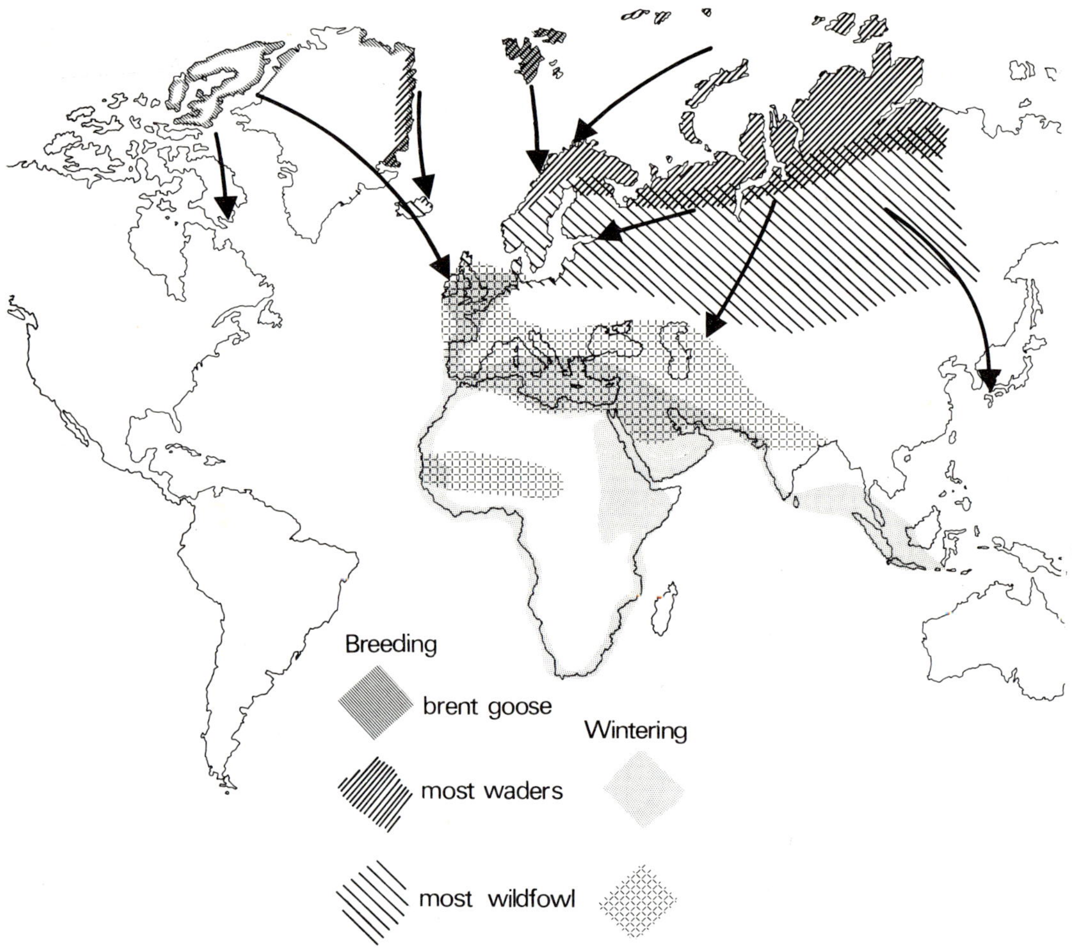

Fig. 8:1. Map to show the main breeding and wintering areas of wader and wildfowl species common on the Bull Island. Some species overlap to a greater extent than is shown. Curlew, redshank and black-tailed godwit nest through Central Europe and Asia: oyster catcher and shelduck only on coasts of N. W. Europe. Most of the wildfowl and some of the wader species also have populations in North America.

We can examine the seasonal population trends from monthly wildfowl counts made over the past few years (*Fig. 8:2*). Teal and shoveler appear first, from August onwards, while wigeon and pintail arrive from early September. Thence forward there is a steady build-up in numbers of all species, with shelduck returning from late October and the first brent geese arriving shortly afterwards. Sometimes there is a bonus such as a white-fronted goose or a pink-foot or perhaps a few gadwall. Normally such birds soon move on.

November is usually the month with the largest number of wigeon and this seems due to the main bulk of Scandinavian and Russian birds passing through to winter further south. Shoveler also reach their peak numbers in November but pintail, though they have sometimes been especially numerous then, do not show the recurring peak of the wigeon. In December duck numbers, except for teal, are slightly lower but in January they increase again with further immigration from Britain and the Continent. Up to 2,500 wigeon, 1,200 teal, 400 pintail and 200 shoveler can be seen in early January, but the pintail and shoveler depart quiet rapidly in late January and early February. Some pintail move to the adjoining estuary at Baldoyle but where the other birds go is unknown at present. It is certainly too early for a return migration to their breeding areas. Wigeon and teal also leave in quite large numbers and the brent geese begin to travel farther afield – to Baldoyle, Clontarf and Sandymount – in search of food. Shelduck depart gradually from February onwards but there is a big decline in April, and by mid-May there are less than 100 on the Bull. Brent leave quickly at the end of April, though a few remain well into May. The only breeding wildfowl in the area appear to be a few pairs each of mallard and shelduck, a situation similar to that mentioned by Kennedy in 1953.

Changes in status

Kennedy published very little quantitative data in his book and this makes a comparison of the wildfowl numbers in the years from 1930 to 1950 with those to-day very difficult. However a comparison can readily be made with the 1950's for detailed counts were carried out from 1951/57. These were not published at the time and have only recently been used for comparison with current numbers (*Fig. 8:3*). It seems that teal, shoveler and pintail have increased in the past 25 years but that wigeon have fluctuated with no long-term change. Shelduck, after an increase in the 1960's, have recently decreased. The growth of teal numbers has been quite marked since 1965, about the time of the construction of the causeway. It is possible that the spread of the plant, *Salicornia*, provided an additional food source attracting more birds. While pintail and shoveler are not so abundant as teal, their increase has been much more obvious and in line with an increase in the rest of the country during the present century. Shoveler were quite scarce prior to 1952, occurring only in twos and threes. To-day 300 may be seen quite often.

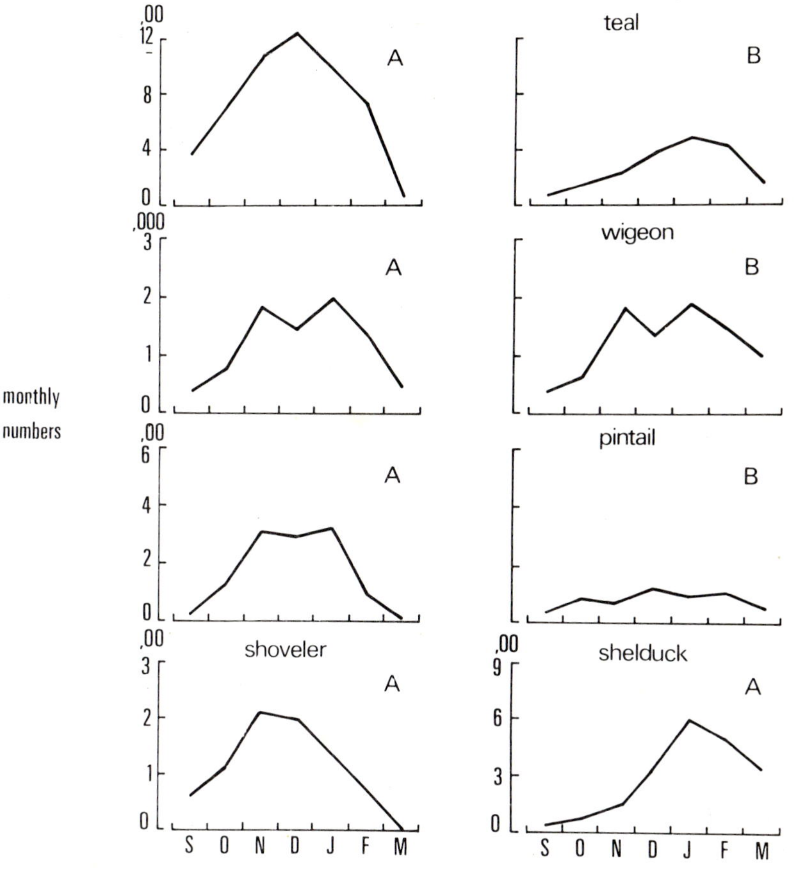

Fig. 8:2. Mean monthly numbers of selected wildfowl species counted at the North Bull and comparative figures for three species.
A 1967-68 to 1972-73.
B 1951-52 to 1957-58.

The fluctuations in the population of brent geese wintering at the North Bull illustrate best of all the fact that the numbers of birds present are dependent on more than local factors such as the lack of disturbance and an abundant food supply. Kennedy (1953) believed that there was an increase after 1930, up till when a flock of 200 birds was considered large. In the early 1950's most counts were of less than 250 birds; figures of 603 in January 1956 and over 1,000 in 1959 were unusually high. From then on there were rarely over 300 until 1965/66 when 460 were counted. Numbers have increased in each subsequent winter until the record count of 1,417 in February 1974. This increase accords with the results elsewhere in Ireland as recorded by the *Wetlands Enquiry* of the Irish Wildbird Conservancy. The proportion of young birds counted in these wintering flocks was high in each year up to 1974/75, indicating continued successful breeding seasons.

Sea ducks and divers

Geese and dabbling duck are certainly the most conspicuous birds on the North Bull. Few people go down to the Point or look out to sea for duck, yet small and relatively inconspicuous diving duck spend the winter there. Flocks of red-breasted mergansers are to be found throughout the winter in numbers up to 30 or 40; goldeneye appear in small numbers (up to 15) in November and remain to early April. Both species probably commute to the Clontarf area where numbers tend to be much higher due to the larger area of feeding available there. Scaup, which occur singly or in pairs off the Bull in most years, probably do likewise, but long-tailed duck, when present, are found chiefly on the north side of Dublin Bay off the North Bull. Both these species may well have become more frequent visitors since Kennedy's time, but the lack of numerical details in the intervening years makes it impossible to know

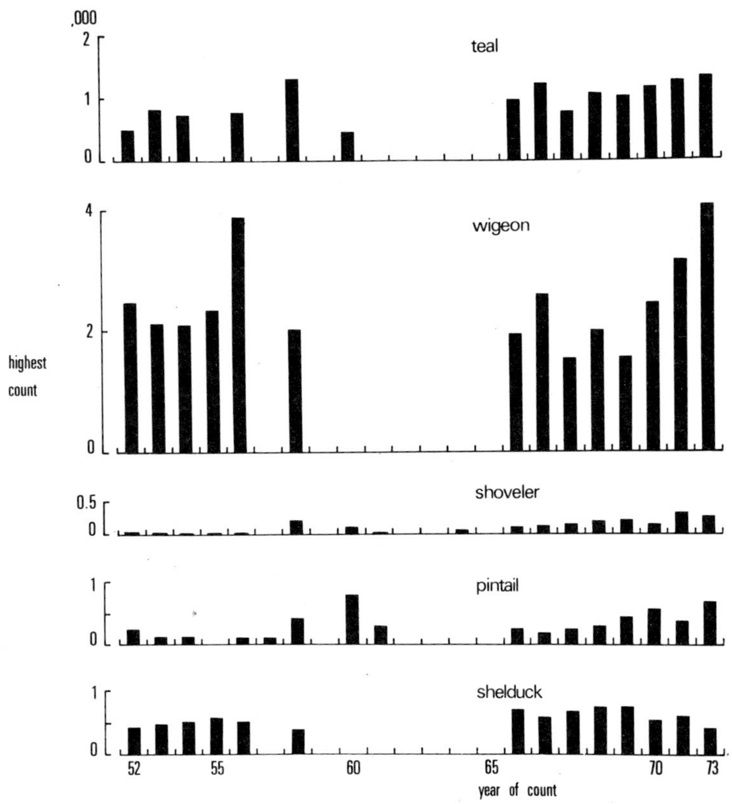

Fig. 8:3. Highest counts of selected wildfowl species at the North Bull in all winters from 1951/52 to 1972/73 in which counts were made in two of the three months from November to January (Teal, Wigeon, Shoveler, Pintail) or from December to February (Shelduck).

whether 11 long-tailed duck in October 1967 and 9 in November 1969 were normal numbers. They have certainly not been repeated since then, but winters have been unusually mild. Out to sea, there is a small wintering population of common scoters off Red Rocks, Howth, from which individuals occasionally occur off the Point with the regular great crested grebes.

The best place for divers is the open sea off the strand, and both common species, great northern, and more especially, red-throated, are regular in spring and autumn, and probably so in winter. A flock of 18 red-throated divers was seen in April 1969 and this size may be regular each spring and autumn when the birds are scattered throughout Dublin Bay.

Waders

The most numerous group of birds on the Bull are the waders. On a typical winter day the casual observer on the saltmarsh will be struck by the great clouds of knot, dunlin *(Plate 8:2)* and bar-tailed godwit which rise in front of him to wheel out over the water and flash the white of their bellies to the sun in unison: he will notice too the shrill teetering of the greenshank and the melancholy piping of the oystercatcher. Aesthetically the waders of the North Bull are immensely satisfying and from the point of view of conservation, they are most important. They occur in huge flocks and for many years this factor more than any other prevented their quantitative survey. As well as this it was felt that the wildfowl were under greater threat so that their populations should be assessed first. The wildfowl counts were done at low tide with the birds spaced out on the mudflats but under these conditions waders are so mobile and dispersed that they are impossible to census accurately. For this reason a programme of wader counts at high tide was initiated in 1971 and counts were made in 33 of the 36 months from July 1971 to June 1974, and again in winter 1974/75 *(Table 8:1)*. Wildfowl were still counted but since they are more difficult to ennumerate in these conditions, the results were not so accurate as before. The main purpose of this programme was to quantify the status of each of the commoner species of wader through the year and, because these counts were only part of a countrywide survey of waterfowl numbers, to establish the national importance of the area as a wader haunt.

In considering the North Bull as a habitat for waders it is important to realise that the island and its surrounding mudflats form only part of Dublin Bay. The birds spread out over much of the intertidal zone when the tide recedes and they may feed anywhere between Howth Head and Dalkey. They also visit Baldoyle estuary sometimes. Within the bay the nature of the feeding grounds varies. The north side and the banks between the Bull Wall and the South Wall are formed of soft mud on which knot and dunlin characteristically feed. The south side is much sandier with frequent cockles and oyster-catcher, ringed plover and sanderling *(Plate 8:3)* feed here. Curlew, bar-tailed godwit and redshank range widely over the bay. At high tide

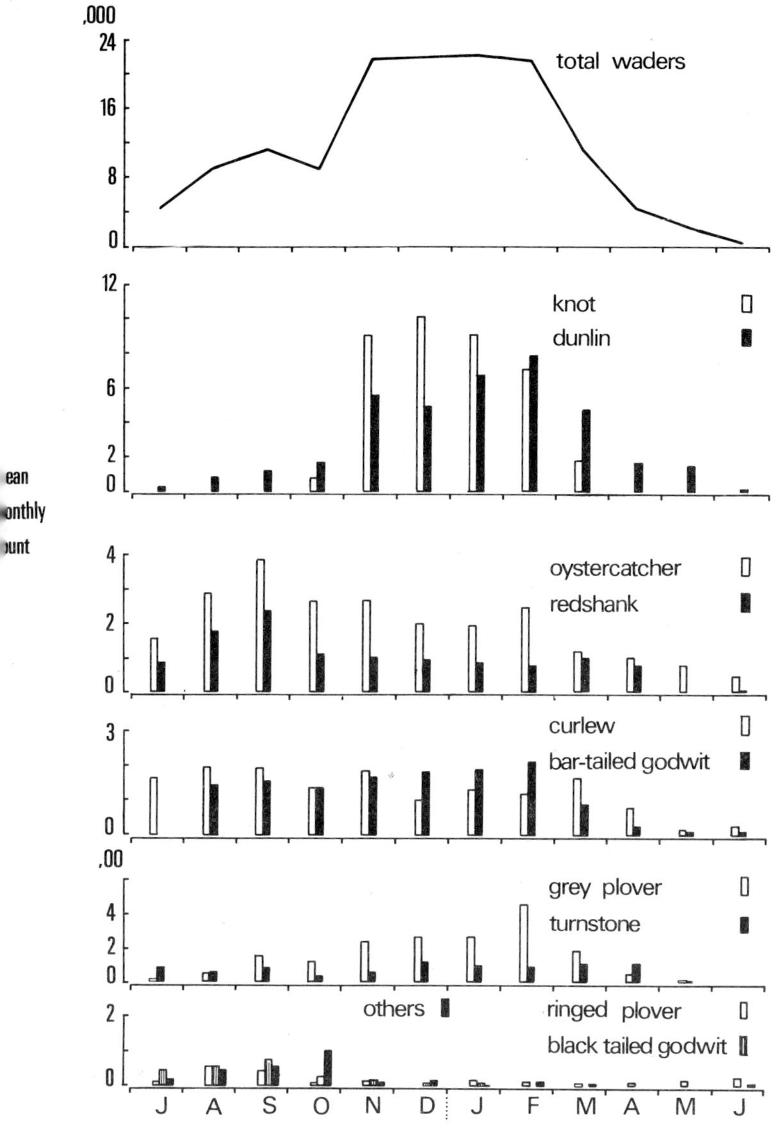

Fig. 8:4. Mean numbers of waders at the roosts on Bull Island by month (July 1971 – June 1974). Peak populations of various species are seen to occur in Winter, in Autumn or occasionally in early Spring. Note differences in scales.

almost all the waders fly to the saltmarsh on the North Bull to roost. Only some oystercatcher (50–100), ringed plover (20–100), redshank (20–150), dunlin (up to 350) and sanderling (up to 200) roost on the south side while a scattering of turnstone *(Plate 8:4)* and purple sandpiper choose rocks and piers around the bay.

Consequently, counts of the wader roosts on the North Bull provide figures for almost all the birds feeding in Dublin Bay. At the same time they emphasise the real significance of the island which is as a safe roost for all the waterfowl. There are four main roosts on the saltmarsh. One long one extends for most of the way between the Bull Bridge and the causeway while three others are situated at intervals north of the causeway.

As a general rule the waders that visit the island nest to the north of the wildfowl but winter to the south (*Fig. 8:1*). Practically all our wader species reach the African continent in winter occurring in large numbers in West Africa with some even reaching the Cape of Good Hope. They make this immense flight in stages so that many birds pass along our coasts in autumn and spring en route to their wintering or breeding areas. These passage migrants spend a few days or weeks in Dublin Bay which is one of three major feeding places on the east coast. The others are Strangford Lough and Dundalk Bay. Some species, like curlew sand-piper, little stint and ruff are almost exclusively birds of passage but in other species only a proportion move on. Other individuals arrive to spend the winter here, though the populations overlap to such an extent that they are most difficult to separate. These species show a peak count in autumn and a smaller one in spring, for example curlew, redshank and oystercatcher *(Fig. 8:4)*. A third group of species, including knot and dunlin feed on our shores mainly in winter. They are the most numerous birds and cause the peak wader numbers to be reached then.

The highest count made at the North Bull so far has been 24,000 waders. Higher numbers have been recorded on the Shannon Estuary, Dundalk Bay and Strangford Lough *(Table 8:2)* but if these are related to the mudflat feeding area it is seen that Dublin Bay supports the greatest density of birds. The International Waterfowl Research Bureau has suggested that a suitable criterion for the listing of a wetland as of international significance for waders is a regular total of 20,000 or more waders. The North Bull clearly satisfies this criterion, whereas it only fulfils the comparable quota for wildfowl with one goose species, the brent goose, and with no duck species. In other words, the North Bull is of much greater importance as a wader haunt than as a duck haunt.

The average monthly populations of wader species in the 1971/74 period is shown in *Fig. 8: 4*. It can be seen that there are large wintering numbers of oystercatchers, bar-tailed godwits, knots and dunlin. In each case the numbers concerned are over 10% of the estimated winter population in the Republic of Ireland.

Plate 8:1. *Wigeon.* One of the commonest and most beautiful of the wildfowl at the North Bull. The relatively small, sharp-pointed bill is adapted for nipping off vegetation and the species feeds exclusively by grazing. Its whistling call is characteristic of the Bull by day and night. *Enteromorpha* is the principle food but the bird also grazes on the salt marsh around high tide. Most of our wigeon nest in Iceland but there is also an immigration of North European birds.

Photo: Richard Mills

Plate 8:2 *Dunlin.* Few who have seen the enormous flocks of waders that congregate and feed on the North Bull can doubt the richness and productivity of the environment. The mudflats may support a feeding flock at this density for up to half an hour and even then the food supply is not exhausted. Dublin Bay attracts the greatest number of waders for its size of any estuary in Ireland and probably also in Britain. Note the unity of purpose among both the flying birds and those on the ground. They must retain this if they are to feed efficiently in such mutual proximity.

Photo: Richard Mills

Plate 8:3. *Sanderling.* A small white wader characteristic of the open beach. It usually feeds on animals disturbed by the breaking waves, following their advance and retreat at a run and picking up invertebrates from the surface. The random nature of feeding contrasts with a flock of knot or dunlin which probe into the mud and locate their prey by touch.

Photo: Richard Mills

Plate 8:4. *Turnstone.* A richly marked wader, the turnstone feeds in small parties on seaweed-covered rocky shores. It lifts up bits of seaweed and small stones ready to snap up crustaceans or insects that are resting below. It has been seen rolling up sheets of *Enteromorpha* on the mudflats. It nests in the high Arctic in Greenland and N. Europe but non-breeding birds often summer on our shores.

Photo: Tom Keogh

These four species occur in large numbers also on Dundalk Bay and Strangford Lough, estuaries which support a wader population very similar to that of the North Bull. The Shannon estuary, by contrast, has proportionately huge numbers of dunlin. The mid-winter numbers of oystercatchers are higher than those found at any other Irish estuary except Dundalk Bay, but twice as many may occur in the autumn for short periods. The numbers of knots are higher than at any other estuary except Strangford Lough. Dunlin are almost as numerous as knot, but they are generally much more abundant throughout the country. The bar-tailed godwit is the common godwit species, as elsewhere on the east coast of Ireland, but black-tailed godwits occur on autumn passage. Redshanks are abundant in autumn when long lines can be seen snaking over the causeway as the tide recedes. In winter there are less than 1,000 and the species is almost completely absent in June and July. Curlews are always present and are commonest from July to March. Reference to *Fig 8: 4* will reveal that two of the most abundant wintering wader species in Ireland, the lapwing and golden plover, were relatively scarce during the three years of counts, being recorded mainly in autumn. These birds are estuarine feeders in some parts of Ireland but have never been particularly common at the North Bull. However, the past four winters have been relatively mild. In conditions of hard weather in Britain or on the continent these birds are among the first to be noted coming in over the sea to Dublin Bay. Large flocks can assemble at the Bull within a few hours, but they pass on quite quickly when the weather improves.

Unfortunately, we have insufficient historical data on the numbers of these more abundant wader species to make comparisons between numbers occuring in even the recent past and those counted during the period of this survey. It is easier to comment on trends in the occurrence of passage migrants in spring and autumn over the years since they are less numerous and more attention has been paid to them, particularly the rarer species. Kennedy (1953) summarised the status of these birds much more precisely than the common species, so a comparison with the present-day status is possible. It should be remembered, however, that the records of the past 20 years are the product of many more hours in the field by many more ornithologists so apparent changes may not, in fact, be real.

Less frequent species

Little stints, which Kennedy described as of rare occurrence, are still not recorded every year though it appears that the main reason is the difficulty of locating them amongst big autumn flocks of dunlin. Curlew sandpipers, on the other hand, are much more regularly recorded, though numbers vary from year to year. In each of the best years numbers in excess of 100 birds were recorded in autumn, with 200 scattered in parties on the channel in September 1959. Ruffs, which Kennedy suspected were probably annual to Dublin Bay, have proved to be so in numbers of up to half a dozen in autumn. Spotted redshanks appeared to be beginning to turn up regularly as Kennedy was writing, and there are indications that two or three have visited the North Bull in almost every autumn since then. There has been little change in whimbrel and greenshank numbers but grey plover do seem to have increased. Kennedy recorded a large flock of 130 but since then 5–700 birds have been noted at times. As can be seen in *Fig. 8:4* grey plover arrive at the North Bull in early autumn and numbers build up monthly to peak figures in December, January or February, probably a similar pattern to that of former years. The reason for the late winter increase is not known for certain. Ruttledge (1966) considered that 'the little clear evidence of passage migration is for the autumn', but the later winter birds are certainly fresh immigrants, arriving at many east and south coast estuaries irrespective of weather conditions. It seems unlikely that they are moving north to their breeding quarters at this time and such redistribution remains to be explained.

Details of the rarer species of waders which include birds breeding in Europe like the avocet and little ringed plover, as well as American species such as the pectoral and buff-brested sandpiper can be found in the check-list. Such records of vagrants thrown off-course by weather conditions illustrate the exciting discoveries which can be made within a short bus ride of O'Connell Street.

Table 8:1. Maximum counts of waders at the North Bull Island 1971–1974

	Autumn July-October	*Winter* November-February	*Spring* March-May	*Summer* June
1971–72	8,500 September	21,000 January	6,000 March	600
1972–73	11,500 October	22,500 January	18,500 March	No count
1973–74	16,000 September	24,000 December	9,500 March	1,000

Table 8:2. Maximum counts and densities of waders at the four Irish estuaries with highest numbers (1971–1974)

	Shannon Estuary	Dundalk Bay	Strangford Lough	Dublin Bay
Oystercatcher	400	10,000	3,000	5,200
Curlew	3,500	4,800	4,000	2,350
Bar-tailed godwit	550	6,000	1,800	2,600
Redshank	3,600	1,600	2,000	2,600
Knot	5,000	8,000	30,200	11,500
Dunlin	33,200	15,500	10,400	10,500
Total waders	52,000	48,000	45,500	24,000
Density/(birds/ha of mudflat)	5·4	10·8	10·1	12·2

Seabirds

To turn to seabirds is to turn to a much neglected aspect of the avifauna of the North Bull: it seems certain that one or two species absent from the check-list have been overlooked. Shags and cormorants are common off the strand and sometimes off the Point; gannets still feed occasionally off the strand; razorbills and guillemots occur irregularly in the channel and at the Point. But no more little auks have been recorded and, even more strangely, there is no published record of a black guillemot *Cepphus grylle* (Linnaeus), even though the species breeds as close as Howth Head. It seems certain that this species has been overlooked. Fulmars, which Kennedy had not noted with certainty, have since been recorded on a number of occasions, particularly off the strand. This is not surprising since they have nested on Howth Head since 1955.

Great black-backed and herring gulls which nest abundantly at Howth Head, Ireland's Eye and Lambay Island, are common on the Bull throughout the year, but most occur in winter. They seem to come to the mudflats mainly to rest after feeding on local refuse tips. Lesser black-backed gulls are regular in summer but migrate south in winter. Common gulls are frequent except in the breeding season, and black-headed gulls are even more numerous. They too leave the area to nest but are only absent for a six week period from the beginning of June. A little gull, an immature and the first record for the North Bull, was seen in October 1968 in the channel near the causeway. Curiously, there has been only one subsequent record of this species which has increased so dramatically elsewhere on the Irish coast. Common, arctic, sandwich and little terns are common off the North Bull in spring and summer, and, except for the little tern, in autumn also. Little terns bred on the island up to 1923; common terns bred in 1919 and possibly for a few years before. Both species, together with the arctic tern, have nested or attempted to nest in recent years. Little terns have bred with varying success since at least 1967; common terns bred in 1969 and 1972; arctic terns have attempted to breed in most recent years. In view of the disturbance on the North Bull in summer, the survival of the little terns is a remarkable feat. Roseate terns are recorded in autumn in most years. Undoubtedly, the most unusual seabird records at the North Bull, and probably the most unusual records of any birds there, were both of terns picked up dead: a bridled tern in November 1953 and a royal tern in March 1954, both new to Ireland, and the latter the first record for Europe.

Land birds

The birds which remain to be discussed are the birds of prey and all those species which depend on the sand-dunes, the salt-marsh and the grassland for their feeding. Although little attention has been paid to seabirds, these landbirds have received even less. Bird-watchers in general look for the unuusual and they come to the North Bull to see waders and wildfowl. The birds of gardens, which they would see at home, would excite no comment and for this reason they have not entered the literature. For example, Kennedy knew of no record for the blue tit for the island and the only sightings that the present writer knows of are his own. There is therefore little that can be added to his comments on these species. Raptors and owls have had more attention than the smaller birds so it is possible to discuss their status on the Bull with more conviction. Kennedy's statement that 'short-eared owls are a usual feature of the bird life of the North Bull during winter' (Kennedy, 1953) is as true today as when he wrote. Numbers vary considerably from one or two up to eight. They prey on mice and small birds such as meadow pipits. Although long-eared owls still breed in St. Anne's Estate there has been only one recent record on the North Bull (in 1971). The commonest bird of prey is probably the kestrel, two or three of which can be seen on almost any day on the island or in its vicinity. The sparrowhawk is also a regular visitor and one or two merlins are generally to be found in winter. Despite its catastrophic decline as a breeding species in Ireland the peregrine is still of frequent occurrence. Although the Bull is no longer within the feeding range of two eyries a bird winters annually on the island, often to be seen perched on the flagposts marking the holes on the golf-courses.

A quick run-down on the known status of a number of species which have not yet been discussed follows: woodpigeons are common on the salt-marsh; kingfishers have been recorded in recent years by the streams at St. Anne's and at Raheny; skylarks are still abundant; two shore larks, the first record for the island, were reported in late autumn 1966; ravens are occasional winter visitors; magpies, which Kennedy apparently did not know as visitors to the Bull, are regular around the causeway; carrion crows *Corvus corone corone*, which may well interbreed with hooded crows *C. c. cornix* in St. Anne's Estate, have been recorded in many recent winters. The wheatear has ceased to breed; due probably to disturbance as well as a general population decline. The most abundant finch is still the linnet. The Lapland bunting was added to the list of birds recorded on the island in September 1966; the firecrest was added in October 1973 and the great grey shrike in October 1974. These few comments sum up the available information: we know very little about the land birds of the Bull and that little can be conveyed briefly. The recent records of rarer birds are probably due to closer watching of the alder marsh in autumn when a trickle of migrants passes through, from farther north and north-east.

Management & its affect on bird numbers

There is a justification for this lack of knowledge about the land birds of the North Bull and that is the fact that the island is a Bird Sanctuary because of its wealth of wildfowl and waders. It is this that makes it so important, not the fact that whinchats have been recorded while coal tits have not. When the Bird Sanctuary was established in 1931 large numbers of wildfowl, especially wigeon, wintered in the channel and on the saltmarsh. The numbers of waders were considered enormous. Since then teal, pintail, shoveler and shelduck have increased and there can be no doubt that the continued high density of the wildfowl population wintering on the North Bull is primarily due to the status of the island as a Bird Sanctuary. The early departure of so many wildfowl at the beginning of February *(Fig. 8:2)* coincides with the end of the shooting season. Some of the shoveler and many of the pintail simply move away to nearby estuaries at this time, suggesting that security from shooting is one of the factors which attracts duck to the Bull in such numbers. Comparison with the position at nearby Rogerstown estuary reinforces this view The shore waders are not subject to the same shooting pressure in Ireland but they too seem to be attracted to the North Bull by its lack of disturbance, in this case for roosting. Unlike wildfowl which can roost on open stretches of water, if necessary, the waders must have a secure roosting place on land. The North Bull is the only area in Dublin Bay to provide the required freedom from disturbance on a scale large enough to accommodate the many thousands of waders. If disturbance was to increase here, the Bay might suffer a large reduction in the numbers present, despite its feeding quality.

However, the mere lack of disturbance will not maintain the bird populations if the feeding habitat changes. Most of the wildfowl, except for brent geese and shelduck, could move inland to feed but the shore waders are by definition dependent on the coastal habitat. With the spread of industrialisation and increasing pressure to reclaim estuaries the number of alternative sites for displaced waders is likely to be reduced in the future, and reduced much more drastically than the alternative sites for ducks. Hence, it is important to conserve the island and its birds not just for the aesthetic value of having large numbers of birds close to the city centre but because the North Bull is one of a diminishing chain of suitable estuaries on the north-west European coast. This means essentially the continuation of the Sanctuary status of the island, avoiding any undesirable change to the saltmarsh where the birds roost and the restriction of reclamation works on the most productive sections of the feeding grounds in Dublin Bay.

Chapter 9
THE ECOLOGY OF WADERS AND WILDFOWL

R. N. Goodwillie *An Foras Forbartha*

We have seen, in the last chapter, the type and number of birds that make the Bull Island their home for some part of the year. Now we must examine their behaviour in order to put them into some sort of perspective within their chosen environment. Each bird must find food in the changing conditions of the estuary; each must find a suitable place to roost and preen. Do they affect each other by competing for food and roosting sites or do they fit into well-defined niches, each separated enough to prevent conflict? They must influence the animals and plants that they prey on but do they take surplus production that would have died anyway or do they actively curtail the growth of these organisms? It is with matters such as these that ecology, in the true meaning of the word, is concerned. The study of them must include a knowledge of bird behaviour, of the life-cycle and habits of each prey organism and of the physical conditions of substrate, tide and weather. It is an enormous subject whose study has only just begun.

General Features of the Estuary

The estuarine habitat is a difficult one for an organism to cope with because of its constantly varying character. At the same time, however, nutrients are abundant in chemical or biological form. Any organism that can adapt to the physical constraints has an enormous food source available to it and so, if successful, it becomes abundant. The habitat is thus characterised by the abundance of a very limited range of organisms. Food chains, from prey to predator are few and short and therefore energy flow is inefficient (see *Ch. 6*). Organisms most often die of old age rather than providing food for a higher layer of consumers. Because of this open-ended system, itinerant species such as birds and fish visit the habitat at various times of the year and take off some of the excess food production. Their presence in such large numbers indicates the prodigal nature of the ecosystem.

At first sight it might be thought that so much food is available to the birds that they would begin at one end and eat their way through to the other, the larger species taking most of the food and the smaller, the pickings. However it is a fundamental rule of ecology that no two species that occur together have identical feeding habits for any length of time. Each has a distinct niche and does not suffer competition except from others of its own kind. It is held within these definable limits either by instinctive behaviour or by its structure. Since the types of food organism available to the birds on the mudflats are limited, different species must often feed on the same food. The mechanism by which each bird exploits the food source without undue overlap is one of the most fascinating aspects of the subject. There are several ways for this separation to occur. Birds could, for example, take prey of different sizes or those at different depths: they could feed at different times or choose different places in the same general area.

Daily Cycle

A bird feeding on the inter-tidal flats must respond to the state of the tide rather than to the 24-hour clock if it is to take maximum advantage of the available food. Therefore its day is split up into roosting during the high tide period and feeding when the tide recedes. Most, if not all of the species have been recorded feeding at night, especially in the winter when daylight is at its minimum. Redshank, for example, seem to feed every night in the depths of winter but only on bright nights in October and March when a longer day allows them more adequate daytime feeding. Generally the species seem to be feeding as they would during daylight but such observations are very difficult as the observer has to have enough light to see the birds in question. Their success, however, is much lower at night and waders usually have only had 1 – 40% of the quantity of food in their stomachs that is found after daylight feeding. Wildfowl are

Plate 9:1. *Dunlin.* A dunlin feeding in as deep water as possible, intent on the movements of prey on the mud surface. When the tide retreats such give-away movements mainly cease and the bird begins to probe. Examination of the bill and wing feathers could indicate where this bird nested and how old it is.

Photo: Richard Mills.

Plate 9:2. *Brent Geese.* A flock of brent geese passing over the causeway on a full tide. Such a scene shows the mixture of beauty and dereliction that is characteristic of most accessible mud flats – the dumping ground of man and sea. The geese have been disturbed from their roost on the northern part of the island and are using up food reserves that have taken time to accumulate. Unnecessary flights like this can result in mortality when the feeding is poor.

Photo: Richard Mills.

well known to feed at night, especially the herbivorous ones like geese and wigeon which depend on a very large throughput of vegetation to obtain sufficient nutriment. One study, in which 76% of the wigeon shot in the morning contained food as opposed to only 21% of those taken in the evening, suggests that night feeding is sometimes more important for this species. It could be of similar importance also to brent geese and shelduck which are species that utilise an abundant food source and can find it by touch rather than sight. Night feeding is thus one possible line of separation that can act between birds which otherwise overlap.

When the wildfowl or waders go to roost they form dense flocks in which adjacent birds would touch each other if they turned round (*Plate VII*). The function of this behaviour would seem to be protective as it allows each bird more rest. Others will be alert to danger and can warn the whole group. There is no social organisation here and it depends solely on the insecurity of individuals on the outside of the flock. They do not feel as safe as if they were surrounded and camouflaged within the group so they remain more alert. The evidence on predation shows that their feelings are justified as predators do concentrate on the outside of a flock. There is always pressure to bunch up further and birds that arrive later tend to drop into thin places inside a flock if they can.

The shorebirds prefer to roost adjacent to water and as the tide rises they are slow to leave each successive site. Wildfowl are apt to roost temporarily on the water, especially if disturbed, though they do go onto land when the tide is full. On spring tides all the species are forced to the back of the saltmarsh but when tides are lower they may line the edges of the channel, in roughly distinct species groups.

As the water recedes the birds become active and begin to preen and stretch. Some leave their roosts almost immediately: redshank, wigeon and teal begin to feed on the saltmarsh and spend only 10% of their time roosting. Others such as knot, curlew and godwits are inactive for much longer periods. The last two species may spend 5 hours roosting around every high tide. This suggests that they are more efficient at feeding than the others, though, being larger, they require relatively less food for their size.

Availability of prey

The time and place in which a bird feeds and its success rate depend largely on the availability of its prey. All the mudflat fauna have a daily cycle related to the tide and so are more easily caught at some times than at others. They may be exploited by different bird species at different tidal phases. The mudflat flora, being static, is available in any period of low water and does not offer the same scope for the diversification of feeding methods. It may be no coincidence that only two bird species depend on it, though its low biomass in late winter and poor nutritional quality may also play a part in this.

Corophium and *Hydrobia* are only active on the mud surface while a film of water remains. On normal mudflats this is for about 3 hours after high tide but in view of the abundance of algal mats at the Bull Island for much of the year, it may be slightly longer than this. In its active period *Corophium* lurks in its burrow, feeding by filtration but also poking its long forelimbs out onto the surface to feel for detrital matter. It seems that it is this action which gives away its presence to a predator such as a redhsank or dunlin. *Corophium* itself responds to such predation: thus either pressure or vibration on the mud reduces the frequency of these feeding movements. In rain for instance, the animal appears less often and the capture rate by birds may be almost halved. This defensive reaction on the part of the prey, which withdraws into its burrow as in the normal resting phase, severely limits the feeding of a flock of birds. It seems to be one of the main reasons why redshank, which specialise on this prey, feed in dispersed flocks. Only at night, when they feed by probing rather than pecking, do redshank bunch together and it appears that they are then taking *Hydrobia* and locating it by touch.

Fig. 9:1. The drawing compares the burrow depth of invertebrate prey and the bill length of wading birds. It is clear that some prey will only be available when they partially emerge from the burrow, hence the importance of pattern in feeding behaviour. Note that invertebrates are not drawn to scale.

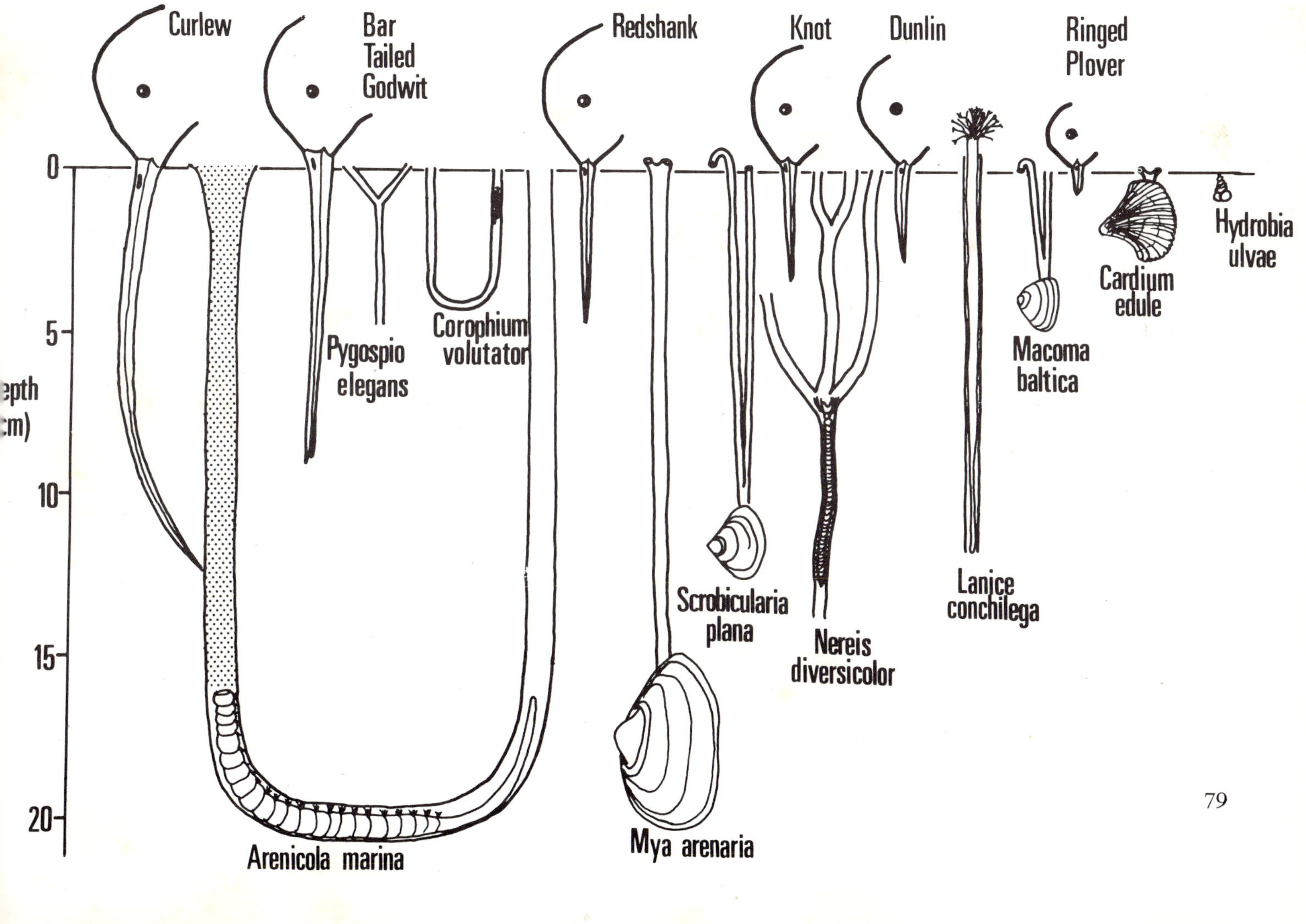

XI. *Salicornia* meadow illustrating the density of cover and extent of colonisation in the N. E. lagoon. *(C. H. Holland)*

Hydrobia itself is probably the major food organism occurring on the mudflats. Almost every species of bird has been noted feeding on it (*Table 9:1*) so its daily behaviour is important. During the ebb tide and for about 3 hours afterwards these snails graze the surface skin of algae that covers all undisturbed mudflats. They leave a distinctive winding trail and are visible to any predator. At this time they are picked off the mud by dunlin, redshank, grey plover and shelduck. This duck eats very little else and must harvest a considerable number of *Hydrobia* every day to extract sufficient food from them. One bird was found with 3,000 snails in its stomach. Its feeding method is to move slowly forward, sweeping its bill from side to side.

Hydrobia begins to burrow when the surface mud dries and it continues to graze underground. Here it is available to probing birds such as knot and dunlin which arrive in hordes from their roost or from *Corophium* areas higher up the mudflats. The shelduck feed more sporadically now, dabbling and sieving in the semi-liquid mud. They await the return of the tide for when shallow water covers the surface once more *Hydrobia* reappears, together with other prey organisms such as *Corophium* and *Nereis*. The shelduck and pintail feed vigorously now, dipping and upending to reach the mud. Since the water is deepening all the time, the wildfowl are left with a clear feeding advantage as the waders are driven shorewards. The redshank, godwits and

Bird species \ Food organism	*Arenicola* (adult)	Deeper molluscs (*Scrobicularia*, *Tellina*, adult *Macoma Mya*)	Beach Annelida	*Nereis*	Fish and shrimps	*Cardium*	*Mytilus*	*Carcinus*	*Hydrobia*	*Corophium*	*Arenicola* (young)	Beach Amphipoda	*Macoma* (young)	*Peloscolex* etc.	*Leptocythere*	Plant seeds	*Zostera*	*Enteromorpha*
Curlew	★★	★★		★				★										
Black-tailed Godwit	★★	★	★										★					
Bar-tailed Godwit	★★	★	★		★													
†*Oystercatcher*		★				★★	★		★									
Grey Plover				★					★★	★								
Redshank			★						★★	★★	?							
Knot							★		★★				★★					
Dunlin				★					★★	★	?		★					
Sanderling			★									★★						
Shelduck									★★	★								
Pintail				★	★			★	★	★						★		
Mallard				★	★			★		★						★		
Shoveler									★	★				?	?			
Teal									★	★						★★		
Wigeon																	★	★★
Brent Goose																	★★	★★

†Oystercatcher is now feeding inland on worms to some extent.

Table 9:1. Suggested major foods of the common wader and wildfowl species at Bull Island

XII. A small creek in the middle marsh. The meander is typically deeper on the outer side, with steep and under-cut banks. The presence of mudflat animals, *Corophium* and *Arenicola* is deduced from the burrow. Lower marsh plants, *Spartina anglica* and *Puccinellia* also follow the creek. Note the large grey bushes of *Halimione* typically fringing the creek. *(C. H. Holland)*

XI C. H. Holland
XII B. Healy

XIII J. J. Moore XIV J. P. Newell

dunlin *(Plate 9:1)* are naturally reluctant to leave a situation where more food is constantly appearing and make the maximum use of their long legs, even swimming occasionally (redshank).

Two of the main predators of *Hydrobia,* the knot and dunlin, form very dense feeding flocks *(Plate 8:2).* However this habit does not significantly affect their feeding success as the animal burrows slowly, if indeed it is aware of the presence of the birds. The stimulus for a flock to move on, which may come every half hour or so, seems to be the result of part of the flock meeting an unfavourable area, rather than when the mud is riddled with holes. The mass movement that follows may involve 5,000 birds and has something of the character of a display flight, though its function is probably observational. As well as this a bird which has such a pronounced loyalty to the flock must very seldom decide to take independent action, so leaving the decision to land to a mutual consensus. When the flock does come down the birds are separated by about 20 cm and begin to feed immediately. They gradually spread out to be about 2 m apart.

Methods of study

Straight observation can tell much about how a bird feeds and what it eats but, because some birds swallow their prey when their bills are still in the ground and some of the prey is too small to identify, it must be coupled with stomach analysis to produce a greater degree of certainty. Dunlin, for example, have three well defined feeding methods called for convenience pecking, jabbing and probing. In pecking the prey is taken from the surface of the mud. They use this method when the mudflats are first exposed, probably taking *Corophium* and *Hydrobia.* In jabbing the bill is inserted half way or less – the depth of many *Hydrobia* as they burrow at low water. Probing to the full length of the bill is used to reach small *Nereis* in their burrows and also filter-feeding molluscs such as *Macoma* and *Cardium.*

As may be seen in *Fig. 9:1* the size and structure of birds offers suggestive evidence of their feeding habits. A longer bill gives the bird access to a greater range of foods; longer legs allow feeding in deeper water. These modifications are linked to some extent to preserve balance in flight and on the land. Between species there are obvious differences but observation is required to show if bill length is really an important factor in feeding ecology. Redshank and greenshank, for instance, seldom seem to probe on the mudflats and their bills may be adapted to their summer feeding behaviour.

There are also sometimes differences within a species: dunlins which breed in Greenland and Siberia have longer bills than our native race. In the bar-tailed godwit the female is noticeably larger and has a longer bill than the male. It generally feeds in shallow water while the male feeds on the adjacent mud surface.

Stomach analysis reveals what a bird has just been eating but its results must be treated with caution. It gives undue emphasis to the molluscs for their hard shells remain in the stomach much longer than the soft bodies of *Arenicola, Nereis* or *Corophium.* In fact they may have a function in breaking up incoming matter. For this reason, only those shells that are intact are sometimes counted because they have a more recent origin. Although this may indicate the absolute numbers consumed there is no knowing the most important food without analysing the prey itself for its food content. Shellfish have seemed to be the major food of such birds as curlews and godwits in some studies. They can make up 80% of the total volume but the evidence, for bar-tailed godwits at least, is that the birds get more food value from *Arenicola* and *Nereis:* species whose only hard parts are minute bristles and jaws.

The time of sampling is also important to note because several birds change their prey as the tide ebbs. A single stomach analysis may conceal the true inter-relationships between species. Dunlin and redshank may feed exclusively on *Corophium* when the upper mudflats are first exposed but dunlin later move down to concentrate on *Hydrobia.* At low spring tides, early in the year, knot may take quantities of mussel spat since the beds are well exposed. But by the autumn most of the mussels have grown too large for knot and are only available to oystercatcher.

XIII. Vegetation of the middle marsh, S. W. portion of the island. The sward of *Plantago, Armeria, Triglochin* and *Limonium humile* is scattered with plants of *Halimione.* Note the hare track.
(J. J. Moore)

XIV. A group of hares feeding in the *Juncus* zone at the transition between salt marsh and dune. Groups such as this may be observed early in the year.
(J. P. Newell)

Many waders produce pellets when too much indigestible matter collects in their stomachs. These give a biassed idea of the food, but when combined with dropping analysis they are of more use. Droppings are adequate for most herbivorous birds since the vegetation passes through them in a little altered form. Being without teeth such birds depend on mechanical means to extract food from the plant material and their gizzards are equipped with very fine particles (less than 0.5 mm) to puncture the greatest number of cells. The grazing wildfowl depend on a large intake of food, whose passage through them can take as little as 20 minutes.

Table 9:1 has been constructed to indicate the various feeding strategies adopted by the estuarine birds. The waders fall into two categories, the larger species that feed at depth and the smaller surface feeders; though the oystercatcher belongs to both. The dividing line in this case is at about 4 cm depth. The wildfowl are all surface feeders but are split into animal feeders and plant feeders. By their bill structure wildfowl can gather large quantities of small food particles easily: waders on the other hand must pick up each particle individually. A large wader would not find the return from this in food value sufficient to justify the effort and so it concentrates on larger, but fewer prey some way below the mud surface. The relative proportions of each feeding strategy can be examined as bird-days and it appears that from a total of 4.5 million/yr in Dublin Bay, less than one quarter of the bird days belong to species that feed at depths below 4 cm. This probably gives a good indication of the relative location of the total food resource.

The bird and its prey organisms

The little work that has been done suggests that the wintering birds take only about 5% of the food available to them during this period. One study has produced a figure of 20% but whatever it is, it is clear that growth in the next season is sufficient to build up the food base to its former level. If it did not, the same number of birds could not be supported in the next winter.

Prey numbers therefore seldom seem to be limiting but prey availability is an important factor. In cold weather many of the invertebrates migrate downwards. *Corophium* makes fewer feeding movements while *Hydrobia* burrows more rapidly after being exposed to avoid becoming frozen into the surface. *Macoma* numbers are reduced by 40% in the top 2 cm of mud. Competition between birds is thus accentuated and they must probe more deeply. If they cannot probe or change their prey they may suffer serious mortality, like shelduck do. Bill length becomes of extreme importance for the waders and it sets species apart that normally overlap in their feeding habits *(Fig. 9:1)*. One can see how natural selection could act effectively by this means: any slight difference that improves feeding being preserved and transmitted to the offspring. It may be significant that the waders as a group evolved mostly during the Glacial period when winters were harsher than at present and the feeding area much contracted.

It is in cold weather that feeding conditions become critical. Day length is short, food is less easily obtained and energy consumption by the birds increased. Roosting is no more a luxury but is of vital importance to conserve energy and food. *(Plate 9:2)*. Any unnecessary flight must be avoided, and safety on the ground is essential.

Chapter 10
SALTMARSH: VEGETATION PATTERN AND TRENDS

J. J. Moore & H. O'Reilly *Department of Botany University College*

The shorelines of Europe near river mouths and sheltered from wave action are covered by a very characteristic type of vegetation usually called 'saltmarsh' by english-speaking ecologists. The continental term for these areas is probably more expressive – 'salt-meadows' (Prés salées, Salzwiesen) since they are grassy and often grazed by cattle or sheep. To the seaward of these consolidated areas, the receding tide exposes areas of mud which can extend out several kilometers. These desolate areas are particularly characteristic of the N.W. coast of Europe where the great European rivers bring down vast quantities of silt and mud. They have an atmosphere of their own and a rich literature mirroring the life of their rather dour human inhabitants. Some of the charm of these areas can be captured on the Bull Island where the pattern of the vegetation on the landward side of the island is remarkably similar to that occurring on the continent. *(Fig. 2:2)*.

The present pattern

As we pass from the intertidal flats covered by algae onto more solid land covered by vegetation comprising the flowering plants (Angiospermae), a series of plant communities may be noted, each contributing to the consolidation of the mud.

The first 'land plant' to colonise the mud is *Salicornia* (glasswort), a succulent plant with its stem swollen with water-storage tissue, its leaves reduced to mere vestiges and its 'flowers' sunken within the swollen stem (*Plate X)*. These send out a shy stigma and anthers when ready for pollination. Two species of glasswort occur commonly on the Bull but they seem very similar to the unpracticed eye. The first coloniser of the mud is *Salicornia dolichostachya*. This species has a longer tapering terminal spike and much bigger root system than the other species. *S. europaea* which is found on the edge of the mud and on the lower half of the saltmarsh. The former species has twice as many chromosomes in each cell ($2n=36$) as *S. europaea*.

The *Salicornias* are annual plants which produce abundant seed in Autumn. Thus in late Spring suitable areas of the mud where there is not too much tidal scour are dotted with their tiny seedlings. There is enormous mortality but in the densely vegetated parts up to 500 plants per m^2 may become established (*Plate XI)*.

The *Salicornia* plants slow down the tidal currents in their vicinity, thus allowing silt and fine sand to settle and build up a small mound around the base of each plant. Soon the scars of the first seed leaves are buried under the mud and this enables us to estimate the rate of its deposition. Over 5 mm. of mud deposition per year has been measured in this way to the north of the Causeway. The microscopic algae, especially the diatoms, form a slimy film over this mud and prevent it from being eroded away by the outflowing tide. Each time the tide comes in they are covered by a new film of mud, but since the diatoms can move slowly towards the light, they soon move upwards and cover the fresh mud surface.

As *Salicornia* is an annual plant, it dies when the first winter frosts come and all that remains after winter has passed are a few sorry skeletons of their tougher, woody tissues, mostly battered level to the ground. Any permanent protection of the mud surface lasting over Winter must involve perennial plants. The first of these to come in, once the mud surface has reached about 10 cm. below mean high water mark is the grass *Puccinellia* which sends long stolons creeping over the mud surface. This stage can be seen very clearly just north of the causeway along the edge of the vegetated marsh *(Plate IX)*. Other perennial plants soon follow, the most noticeable being sea lavender, *Limonium humile,* and soon a firm surface is formed.

North Bull Island – Dublin Bay

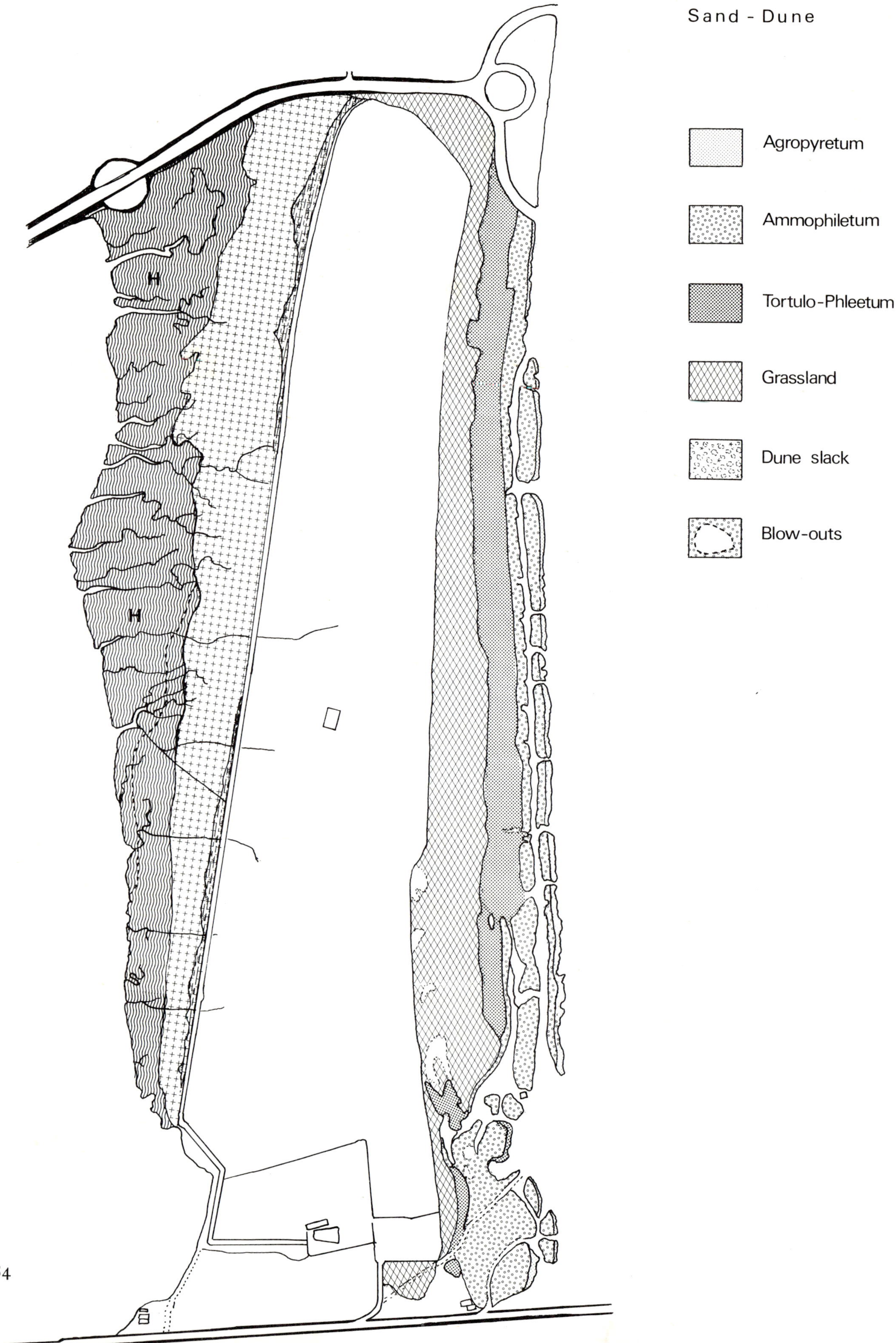

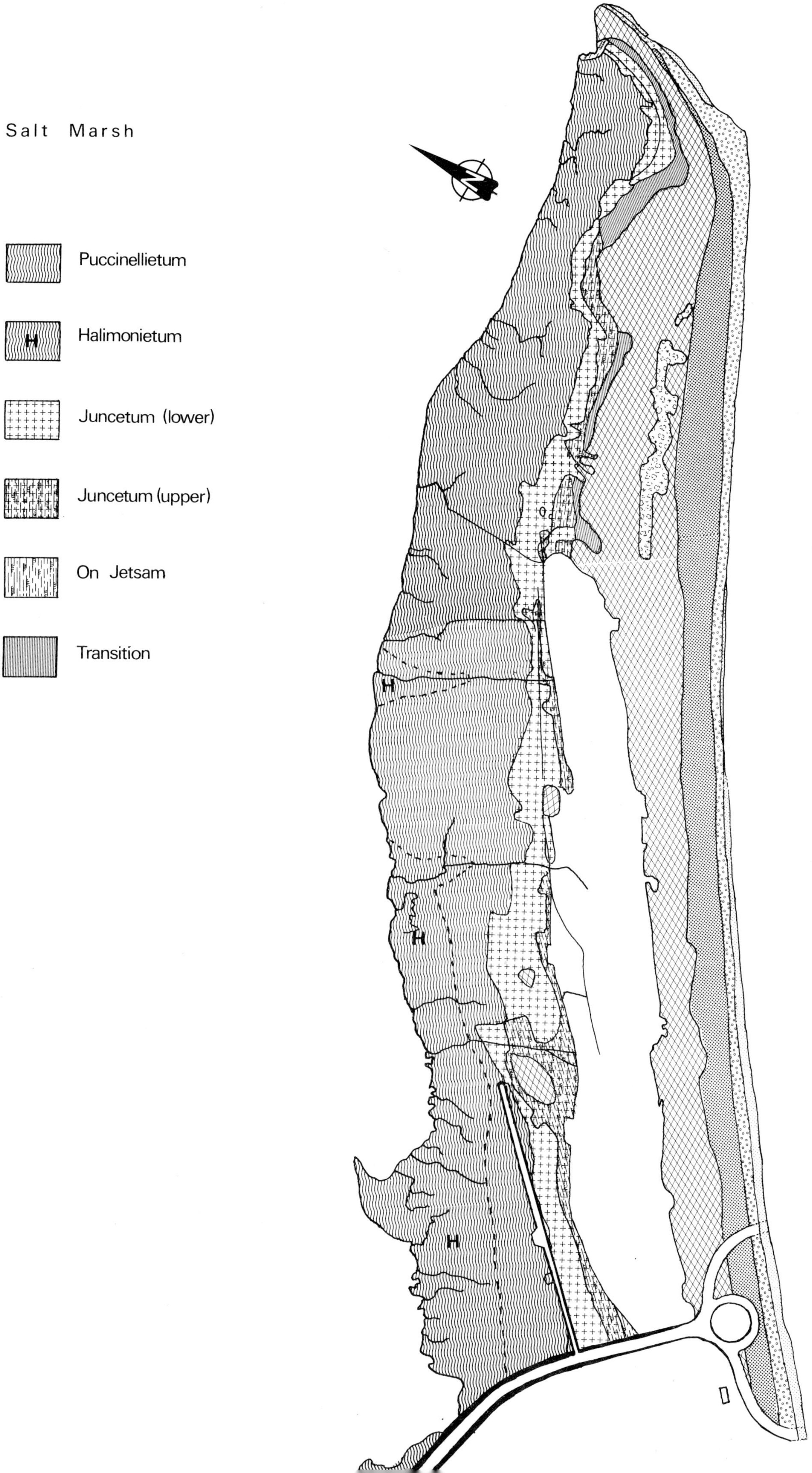
Salt Marsh
Puccinellietum
H
Halimonietum
Juncetum (lower)
Juncetum (upper)
On Jetsam
Transition
H
H
H

Except for this area of mud colonisation most of the front of the marsh along the length of the island is marked by an eroding edge or step which varies from 20 cm. to 60 cm. high *(Plate I)*. This eroding edge was noted in 1951, long before the causeway was built and was presumably caused both by the tidal scour and by wave action when the wind was in the right direction (SW). Behind this eroded edge the grass *Puccinellia maritima* is still very vigorous. It is this grass which gives its name to the vegetation of this zone *(Puccinellietum)* extending from mean high water (MHW) to about halfway up the marsh. There are only two other higher plants strictly confined to the severely fluctuating conditions of this part of the marsh *(Spergularia marginata* – spurry – and *Suaeda maritima* – sea blite) but there are also a few very characteristic algae occurring here such as the green *Enteromorpha* species and the 'fuzzy-haired' red alga *Bostrichia scorpioides*. In addition, the plants which occur right across the marsh soon become established here. There are three species which are widespread in all European salt marshes – sea plantain (*Plantago maritima), sea* aster *(Aster tripolium)* and sea arrow grass *(Triglochin maritima) (Plate XIII)*. The sea pink or thrift *(Armeria maritima)*, which in June clothes the middle part of the Bull marsh in pale pink, occurs further towards the front of the marsh than on the more easterly coastal marshes of Europe.

About halfway up the marsh, 15 cm. below the mean spring high tide level (HWST) a subtle change in the vegetation takes place, not very noticeable to the eye, but correlated with a change in the soil. Species characteristic of the front of the marsh fade out (*Puccinellia, Suaeda* and *Spergularia*) and the first plants that characterise the upper marsh appear, such as Sea milkwort *(Glaux maritima)* and the scurvy 'grass' *(Cochlearia officinalis)* whose tiny white flowers appear scattered in the low vegetation in early Spring. As we move upwards a fine-leaved grass, the red fescue *(Festuca rubra)* also appears. This grass is sometimes confused with *Puccinellia* but has fresh, bright green foliage as against the blue-green of *Puccinellia.*

Further up still, we reach the most obvious zone on the north end of the marsh, the area dominated by the sea rush, *Juncus maritimus (Plate XIV)*. Between the clumps of this large, harsh rush one can find the inconspicuous and softer *Juncus gerardii* which gives its name to the whole of the upper community. The reason the European vegetation ecologists use the name of this tiny rush, which is seldom more than 15 cm. high, rather than that of the larger *J. maritimus,* is that the latter is quite rare in the S.W. European marshes and occurs, if at all, above the inundated zone.

Superimposed on this basic pattern is another community that has only developed recently, the *Halimionetum portulacoidis,* dominated by the shrub, sea purslane *(Plate XIII)*, conspicuous by its white mealy leaves.

As one moves back towards the sand-dunes, especially where the transitional zones are still intact and not disturbed by the golf links, the lush green of *Festuca rubra* dominates the scene. A number of interesting plants occur here. The most notable of these is a second, larger species of sea lavender, *Limonium binervosum,* recognised easily by the two clear veins to each side of the midrib when the leaf is held up to the light. This plant is quite rare in Ireland, though it does occur in abundance on Malahide 'Island', 10 km. further north. Also present, hidden among the dense growth of the red fescue are two species of sedge which occur only in this transitional zone – *Carex extensa* with green fruits and *Carex distans* with brownish fruits.

Thus, as can be seen in the vegetation map *(Fig. 10:1 a & b)*, the pattern that emerges is of a series of zones running more or less parallel to the shore as shown in *table 10:1*

Table 10:1 Saltmarsh Vegetation Units.

1.	*Salicornietum dolichostachyae*	
2a.	*Puccinellietum maritimae;*	*Puccinellia* facies
b.	*Puccinellietum maritimae*	typical sub-association
3a.	*Juncetum gerardii*	typical sub-association
b.	*Juncetum gerardii*	sub-association of *Juncus maritimus*
c.	*Juncetum gerardii*	facies of *Festuca rubra*

The changing pattern

Saltmarshes are very prone to change. The accumulation of silt and clay which is consolidated by the growth of the vegetation means that the ground level is continually rising and thus the time of submergence by the tides becomes less and less. These inherent changes may of course be altered by large-scale subsidence or elevation of the whole coastline.

Saltmarshes with their fine-grained soil material are also very prone to erosion. The pattern of winding drainage channels is a striking feature very obvious from the air *(Plate 2:2)*; these channels may enlarge, especially towards their mouth and sharp eroded edges are a feature of the majority of Irish saltmarshes. The Bull marsh, as we saw, is no exception. All of these changes will be reflected in the vegetation.

In the case of Bull Island we are fortunate in having a series of aerial photographs and also a series of studies on the vegetation, which enable us to follow most of the changes. The first vegetation survey dates from 1951 (Denneny) and good aerial photographs are available since 1954.

The three most remarkable vegetational changes that have taken place since that time are:

1. The spread of *Halimione* and its present vigour in the areas marked on the map. A large part of O'Reilly and Pantin's study, published in 1957, is taken up by the problem of 'the sporadic or very localised occurrence of *Halimione* . . . in contrast to its relative dominance in the Baldoyle area'. It had already been noted as very rare by Denneny in 1951 both north and south of the causeway. At present, in 1976, it cannot be described in any other terms except 'dominant' in the areas marked 'Halimionetum' on the map.

This increase had not been noted north of the causeway in 1969 by Griffin; by 1971 however, it was noted that, in the angle between the causeway and the Dublin Corporation embankment, all of the middle marsh, not just the dykes and channels, carries *Halimione*. Likewise, in 1972, Ó. Ní hAnnracháin recorded the widespread occurrence of the plant to the south of the causeway, as noted in the vegetation map.

2. The dramatic increase in *Salicornia europaea* and *S. dolichostachya* in the mudflats north of the new causeway subsequent to the blockage of tidal flow in 1964. In 1957, O'Reilly and Pantin noted only three patches of *Salicornietum*. Fig. 11:1c shows the extent of *Salicornia*-dominated mudflats in 1975. This spread is part of the natural dynamics of marsh growth when mud accretion is occurring.

3. The explosive spread of *Spartina* in recent years: the story of this interesting and controversial plant deserves a special section to itself and is dealt with in the following chapter.

Acknowledgements
The authors wish to thank Joan Denneny, Phillip Griffin and Orla Ní Annracháin for the use of information obtained in undergraduate projects.

Chapter 11
SPARTINA ON BULL ISLAND

P. J. Boyle *Department of Botany University College*

An account of *Spartina* on the Bull Island forms part of a complex and continuing saga, in which ecological, genetic, ornithological and sociological strands are interwoven. The aim of this account is to indicate the present situation of *Spartina* on the Island as a guide to further study.

The species of *Spartina*, or cordgrasses, are numerous and occur in both freshwater and marine habitats throughout the U.S.A. Elsewhere, species are confined mainly to muddy estuaries and neighbouring salt marshes. *Spartina* was not known in Ireland prior to 1925 when it was introduced from the south of England. Other imports followed and as most were purchased by the Irish Government, they are fairly well documented, but subsequent transplantings are not, so that the precise source of many Irish plantings, including those at the Bull Island, are not known. *Spartina* was exported to many other countries in addition to Ireland. In all cases, the grass was planted with a view to land reclamation and coastal defence against erosion and successful introductions were from offsets.

The grass exported to Ireland and elsewhere was known at the time as *S. townsendii*, or by the common name 'rice grass'. It is now known that the title included grasses currently recognised in England as *S. x townsendii* and *S. anglica* and that exports may have been of one, or other, or both of these. When there is doubt about the exact type of grass in a large plantation, the title *S. x townsendii* agg. is sometimes used.

Two other forms of *Spartina* could have been included inadvertently in exported material, although there is no evidence of this. However, as Irish plants are identified by comparison with plants from the parent source, it should be noted that there are four main forms of *Spartina* in England (Goodman et al. 1969; Hubbard 1968):

(i) *S. maritima*, rare; thought to be indigenous;

(ii) *S. alterniflora*, very rare, or extinct; thought to have been introduced;

(iii) *S. anglica*, very common, male-fertile, seed-producing; thought to be a polyploid of origin from *S. maritima* x *S. alterniflora;* growing plants known in abundance since c. 1890;

(iv) *S. x townsendii*, not uncommon, but scattered, male-sterile, not seed-producing; some plants thought to be diploid hybrids from *S. maritima* x *S. alterniflora*, other plants thought to be haploids from *S. anglica;* growing plants first noted in 1956 (Hubbard 1957).

Though attractive and popular, this hypothesis of hybridity within the genus *Spartina* remains unproven since first suggested in 1908 by Stapf and it should therefore be treated with caution.

Forms of *Spartina* on the Island and how to recognise them.

There are three forms of *Spartina* and a number of variants on, or in the immediate vicinity of the Island (see *Fig. 11:1c* for general location and distribution). Two forms of the grass predominate. They are in general agreement with descriptions of

S. x townsendii and *S. anglica* in Britain (see *Table 11:1*). Distinctions between the two Irish grasses are less clearcut than for British material, partly because an overlap in range of variation reduces the diagnostic value of some traits of the Irish material. The presence of seed is always diagnostic of *S. anglica*. Prior to seed production, the predominance of long bristles and exserted (i.e. protruding and dangling) anthers may be used with reasonable safety to distinguish *S. anglica* from *S. x townsendii* on which short bristles and enclosed anthers predominate. Of course, these features should be examined in conjunction with others.

On bare mud, both *S. x townsendii* and *S. anglica* produce vigorous, horizontally-growing rhizomes and vegetative spread by this means may be considerable in any year, usually giving roughly circular clumps *(Plate VIII)*. Rhizomatous spread is inhibited on gravelly substrata and where *Spartina* is surrounded by general salt marsh vegetation. Rhizomes do not compete with underground parts of other plants but they spread rapidly in any uncolonised muddy surface exposed in general salt marsh vegetation, e.g. newly dug drains, tractor marks, uncolonised mud between bushes of *Halimione* etc. *(Plate XII)*.

Table 11:1. Summary of range in variation of traits of *S. anglica* and *S. x townsendii* in Britain and Ireland.

	BRITAIN		IRELAND	
	S. anglica	*S. x townsendii*	*S. anglica*	*S. x townsendii*
Height cm	30–130	30–130	6–192	22–102
*Ligule mm**	2–3	1–2	1–4	1–3.5
Leaf length (cm)	10–45	6–30	2–46.5	1–42
Leaf width (cm)	0.6–1.5	0.4–1.2	0.2–1.5	0.1–1.1
*Leaf angle in degrees**	widely spreading	20–45	6–90	5–50
Panicle length (cm)	14–40	up to 25	5.5–44	6.8–33
Spike number	2–12	2–9	1–15	2–7
Bristle length (cm)	up to 5	1–2 or more	0–3.5	0–1.6
Spikelet length (cm)	1.4–2.1 hairy	1.2–1.8 hairy	1.3–2.2	1.1–2.1
Spikelet width (mm)	2.5–3.0	2.0–2.5	2–3	1–5
*Anther length (mm)**	8–13	5–8	5–12	3.5–9.5
*Anther behaviour**	normal	abnormal	normal to abnormal	abnormal
*Seed**	present	absent	present but rare to abundant	absent
*2n**	122	62	124	62
*Pollen grains**	normal	abnormal	abnormal and normal in one anther	abnormal

*Indicates features of particular diagnostic importance.

Plate 11:1. Clump of *S. maritima* forma *dublinensis* with foliage of *S. anglica* in the background. Note the compact growth habit. (Ruler graduated in inches).

Plate 11:2. Row of *S. x townsendii* clumps at Bull Wall end of salt marsh, looking north-east in 1965. Note that the clumps are confined to pans in the general salt marsh vegetation.

The third form of *Spartina* is conspicuous by virtue of its dark-green foliage, closely tufted habit, vertical rhizomes and low flower production (*Plate 11:1*). Predominantly upward growth of rhizomes tends to keep clumps of this grass small relative to those of *S. anglica* and *S. x townsendii*. To date, only one clump has been found on the Island, but similar clumps are present in estuaries on the east coast. This grass is closer to *S. maritima* than to any other form of *Spartina,* but is sufficiently distinct to merit the provisional title *S. maritima* forma *dublinensis* (Boyle, in press).

Pending further investigation, other variants of *Spartina* may be included under the name *S. anglica*.

None of the forms of *Spartina* in Ireland has been observed to spread naturally by detached fragments, although artificially uprooted sods may re-establish. *S. anglica* is the only form known to spread by seed.

History and spread of *Spartina* on Bull Island

Evidence indicates that both *S. x townsendii* and *S. anglica* were planted, but the precise source of either grass is not known. New clumps of *S. anglica* and variants including *S. maritima* forma *dublinensis* have originated from seed of *S. anglica* during the past 5–6 years.

S. x townsendii was first recorded on the Island as *S. townsendii* in 1934, when clumps of the grass were 30–50 cm in diameter on bare mud (personal communication from the late Professor Doyle). By 1949, O'Reilly and Pantin noted that the clumps were c. 3 m in diameter, situated in the upper marsh and set more or less in a straight line, indicative of deliberate planting. An investigation commenced by the writer in 1958 revealed that the grass had been planted in at least 43 marsh pans of various sizes and shapes, from 10 x 10 m down to 1 x 1 m. It had spread by rhizomatous growth to fill the pans and form clumps which reflected pan shape and size. All clumps were contained by pans and showed no inclination to spread beyond them (*Plate 11:2, Fig. 11:1a*). Clumps were sufficiently uniform to appear clonal. The grass was male-sterile, seedless and in general agreement with *S. x townsendii*. These features were noted independently by Dr. C. E. Hubbard of Kew in 1957 (personal communication) when he lodged a specimen in the National Herbarium in Dublin.

The Bull Island is the only site in which it is known with certainty that *S. x townsendii* was planted in Ireland; it is a unique stand of the grass and nothing comparable has been recorded from Britain, or elsewhere.

During construction of the causeway, four clumps of *S. x townsendii* were obliterated when the causeway was built over them. By 1970, the central parts of some clumps north-east of the causeway had been colonised by *Halimione,* but otherwise the grass showed little change at that time. Since then, many clumps of *S. x townsendii* have been wiped out, or severely damaged following spread of, and attempts to eradicate *S. anglica*.

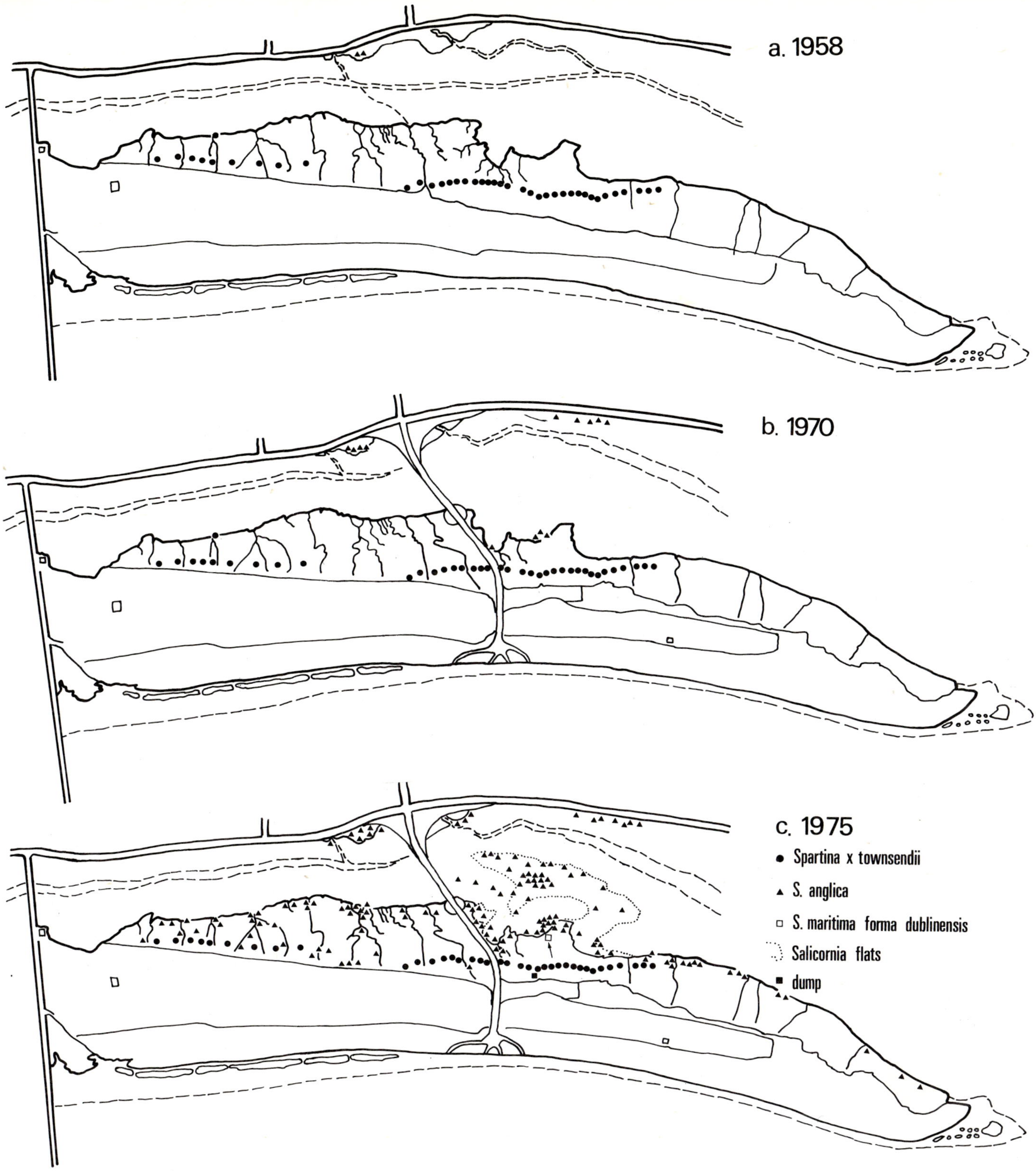

Fig. 11:1. Changing distribution of the three forms of *Spartina* on Bull Island: (a) Pre-causeway (1958); (b) Post-causeway (1970) and (c) Winter, 1975.

The latter species was present at one site only in 1958, when there was a patch about 2 m in diameter at the south-eastern side of Watermill Lane Marsh *(see Fig. 11:1a)*. During causeway construction, the marsh was disturbed, muddy surfaces were exposed, the continuous channel between Island and mainland was interrupted and tidal currents were altered. These factors enabled *S. anglica* to spread both by rhizome growth and by seed. By 1970, there was a large patch surrounded by seedlings at the original site, on mud, south of the causeway entrance; there were 4 – 5 small clumps on muddy gravel north of the causeway entrance; and there were two clumps and associated seedlings on accumulating silt in the angle between the Island and the north-eastern shore of the causeway *(see Fig. 11:1b)*.

Attempts were then made to eradicate *S. anglica;* these included clipping, digging up and treatment with herbicide. However, eradication procedures hastened rather than controlled the spread of the grass so that by Autumn 1975 seedlings were established in areas away from tidal scour on both open mud and salt marsh *(Fig. 11:1c)*. Small clumps are particularly evident and thriving in the vicinity of channels on the Island and in areas of high silting on the mud flats. Innumerable tufts are present among general salt marsh plants and although they are stunted, they are seed-producing and thus a source of more *S. anglica.*

Clipping of some clumps in winter 1971 had little if any effect on growing plants; instead it enabled panicles containing mature seed to spread to haunts congenial for germination.

Most of the digging and all the herbicide treatments were large-scale operations carried out by Dublin Corporation in response to a plea to protect the mud flats from invasion and thus retain them as feeding grounds for wading birds. Unfortunately, the Corporation was not made aware of the uniqueness of *S. x townsendii,* nor of the fact that it was not spreading (personal communication from members of Dublin Corporation). Consequently, in summer 1973, all conspicuous clumps of *Spartina* were dug up and transported by tractor to a dump on the Island (see *Fig. 11:1c* for location of dump). Clumps of *S. x townsendii* farthest from the causeway were damaged most severely, but pans left by its removal are now fringed by tufts which have originated from rhizomes overlooked during digging. Sods were not removed from stands nearest the causeway and these are now fairly well re-established, although contaminated in places by *S. anglica* and other marsh plants. Some large clumps of *S. anglica* were removed, but seedlings and small tufts were unharmed. New *S. anglica* plants have become established in tractor tracks, in muddy pans and in various places where sods were dropped *en route* to the dump.

In 1973 and 1974, *Spartina* on the Island was treated with the herbicide Dalapon, but with little or no success. In August 1975, 'all *Spartina* in sight' was treated with the herbicide Hyvar-X, which is effective against a wide range of annual and perennial plants, including rhizomatous grasses. The effects of spraying will not be evident until the next growing season but it appears that some rhizomes have survived. It is hoped that *S. x townsendii* has not been eradicated by the harsh treatments it has suffered. It is certain that small patches of the seed-producing *S. anglica* will survive all treatments because of their very number.

Further study and possible management

Now that the scientific interest of *S. x townsendii* is recognised, appropriate measures of conservation should be employed.

Further study of the pattern of spread of *S. anglica* is considered to be of great interest. The species is situated in highly dynamic environments, both on the salt marsh and on the mud flats: south of the causeway areas of salt marsh are being eroded, while areas of mud-flat north of the causeway are undergoing considerable siltation. If the areas undergoing siltation are not colonised by *Spartina,* they will be colonised by *Salicornia* and most of them have been so colonised already. In either event, vegetation will develop on the mud. If the vegetation is *Spartina,* it will release detritus at every outgoing tide and this will lead to changes in micro- and meso-fauna on the mud flats. If the vegetation develops towards that of a drier marsh type, it may lead to the elimination of the mud flats. However, the final extent and nature of vegetation on the mud flats is unpredictable. In order to maintain the *status quo,* or something like it, it is necessary to reduce silting north of the causeway. A method of doing this might be the re-introduction of a free-flowing channel of tidal water between the Island and the mainland, i.e. through the causeway. This would both maintain the mud flats and might even arrest the severe erosion of the salt marsh now occurring south of the causeway.

Acknowledgements

My most grateful thanks are due to Mr. R. Billen, B.Sc., for assistance in preparing this account. Thanks are also due to Mr. D. Dooge of the Dublin Naturalists' Field Club as well as to many members of Dublin Corporation for patient and sympathetically-given information. Especial thanks are due to the Irish National Science Council for funding research on *Spartina.*

Chapter 12
SALTMARSH FAUNA

B. Healy *Department of Zoology University College Dublin*

The saltmarsh as a habitat for animals

A saltmarsh represents a much wider range of ecological conditions than either a beach or intertidal flat. Marine, freshwater and terrestrial environments and various intermediate conditions exist side by side in a complicated pattern of microhabitats. In aquatic habitats, salinities may vary from almost fresh to full sea water, water may be temporary or permanent, still or running. Terrestrial habitats may be sandy, peaty or muddy, well drained or covered for long periods by standing water. In addition to these spatial variations there are temporal changes, conditions fluctuating according to tide cycle, rainfall and temperature.

To survive these environmental permutations the animals permanently resident on saltmarshes must have special adaptations, structural, physiological and behavioural; above all they must be tolerant of environmental change. Considering the range of ecological niches available, the number of species which one could expect to find with some degree of certainty at any one time is small when compared with, say, a freshwater marsh or a bog. Among 329 species of invertebrate identified during a 3 year survey of the Bull Island Saltmarsh carried out in 1961–63, only 40 can be described as constant, i.e. permanent members of the saltmarsh community. The remainder are either sporadic, arriving accidentally and surviving only so long as conditions are tolerable; or seasonal, present during a limited period each year. We can conclude that existence on a saltmarsh is hazardous and that few species have been able to adapt to the demanding and often unstable environments.

As in all intertidal habitats, life on the saltmarsh is dominated by the ever present influence of the tides, this influence being greater in winter than in summer. The exposure and salinity gradients resulting from tidal immersions determine the distribution of the animals and give rise to a zonation similar to, but less well defined than that shown by the vegetation. The degree of exposure, that is the length of time a particular part of the shore remains uncovered by tides, affects marine and terrestrial faunas in different ways. Marine organisms generally require more or less regular periods of immersion in sea water for their survival and tend therefore to be concentrated on the lower marsh while terrestrial species avoid immersion as much as possible and are most numerous in the upper zone. Some marine animals, however, are able to withstand long periods of exposure and some terrestrial species are unharmed by occasional immersions. It is these adapted species which are the most important members of the community and which may be described as true saltmarsh species.

Salinity of freestanding or running water is an important factor for all animals. Generally speaking, the lowest salt concentrations are found in the upper zone but heavy rain can lower salinity dramatically in all zones. A very restricted freshwater fauna exists in the rain-filled, shallow pans and depressed areas of the Juncetum near the tip of the island but the amount of water draining from above the marsh is not sufficient to create freshwater conditions in the channels. Full strength sea water may be present for limited periods in the pans and near the outlets of the channels thus true marine species can survive for a while in these situations. Only rarely are salt concentrations higher than sea water encountered in pans or channels. The majority of aquatic habitats, however, are brackish and the fauna consists for the most part of brackish species adapted to low and variable salinity. The way in which soil salinity affects terrestrial animals is not so well understood but the fact that nearly all the common species of freshwater marshes are absent here suggests that salt is an important limiting factor.

The channels and pans

The aquatic habitats of the marsh are characterised by their instability. Water level, salinity, temperature and oxygen levels are all subject to frequent and often rapid fluctuations and all play a part in determining the numbers and distribution of the fauna. The variations are greater in the pans than in the channels and this is reflected in the poorer and more sporadic nature of the pan fauna.

The tendency for water bodies to dry out is probably the greatest hazard for aquatic animals. Pans are frequently dry and standing water in them represents, at most, a temporary habitat. Although most tides enter the channels at high water, neap tides may only be felt near the outlets. Thus if dry weather with high evaporation and reduced water draining from the marsh occurs during the times of neap tides, the water level may fall until only isolated pools remain; the smaller channels may be completely without water cover for extended periods.

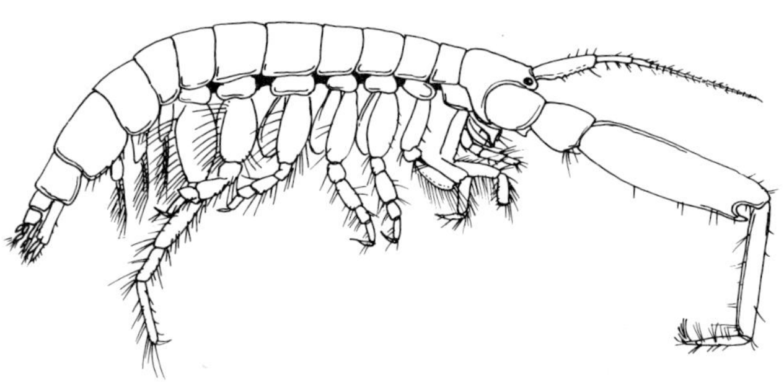

Fig. 12:1 *Corophium volutator,* a burrowing amphipod. (Natural size 4-8 mm.)

The most effective way for animals to avoid desiccation during periods of drought is by burrowing into the substratum or banks and in fact nearly all the common aquatic species of the marsh burrow to some extent. This behaviour is shown not only by worms and bivalves which burrow as a way of life but also by snails, crabs, shrimps and even, in a limited way, by fish. In the case of snails, additional protection from water loss is provided by a shell and operculum and some species e.g. the rough winkle, have physiological adaptations such as the ability to produce a concentrated urine.

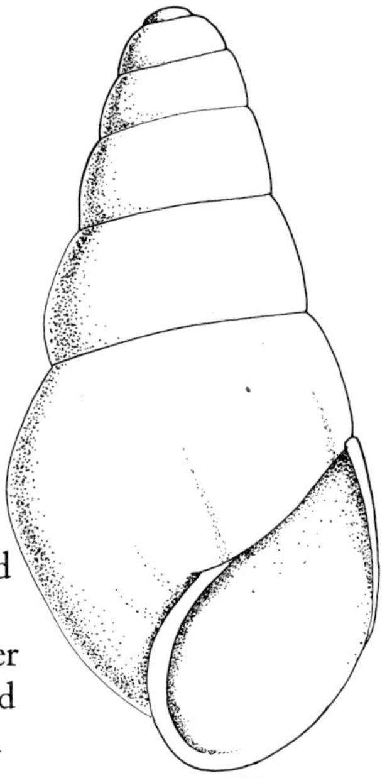

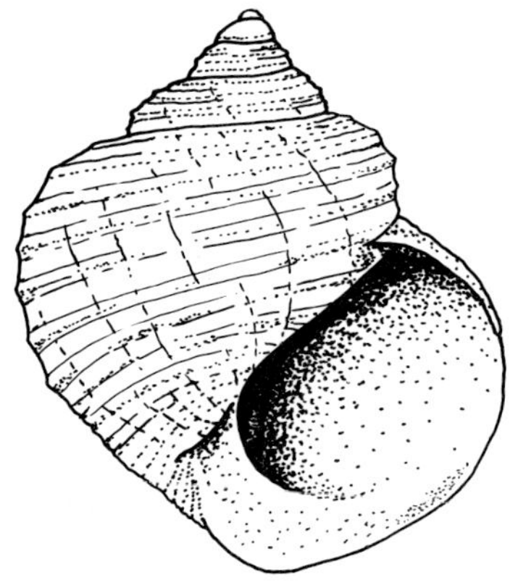

Fig. 12:2 *Hydrobia ulvae,* the mud snail. (Natural size 1-6 mm.)
Fig. 12:4 *Littorina saxatilis,* the rough winkle. (Natural size 5-12 mm.)

The problem posed by variations in salinity concerns the ability of an animal to control the concentration of its body fluids. Marine animals normally have their body fluids (blood or coelomic fluid) at the same concentration as sea water. When such an animal enters water of a lower salinity, water diffuses into the body by osmosis causing it to swell and, if not controlled, causing damage to the tissues. There is also a tendency for salts to be lost from the body. The control of osmotic potential – osmoregulation – can be achieved in several ways and in fact most common sea-shore animals are capable of some degree of control. Very few, however, can adapt to the wide range and often rapid fluctuations which are met on a salt marsh; the brackish species of the marsh are therefore highly specialised animals. Tolerance of low salinity is greater at higher temperatures, which explains why some animals such as the shrimp (*Crangon vulgaris*) are more likely to be found on the marsh in summer. On the other hand, oxygen is less soluble at high temperatures and the combination of low salinity and low oxygen concentration can put a severe strain on an animal's system.

Fig. 12:3 *Carcinus maenas,* the shore crab. (Marsh specimens are all juveniles and vary in size between 1-5 cm.)

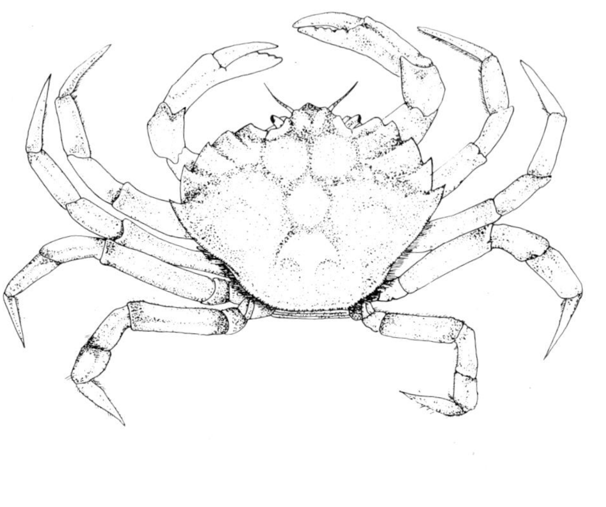

Fauna living in the substratum

The combined effects of exposure, low salinity, high temperature and low oxygen levels affect particularly the upper zones and the result is a distinct zonation of the fauna. In a survey carried out from 1961–1963, 13 species were recorded in channels from the upper zone, 34 from the middle zone and 37 from the lower zone. This zonation underlines the dependance of most species on tidal influence.

Whilst salinity differences of the water body have been emphasised so far, the substratum of pans and channels may be considered as an extension of the mud flats. (*Plates 2:2 & XII*). Most of the common burrowing species of the sand and mud flats penetrate into the salt marsh. The degree of penetration depends on the ability of the species to withstand the environmental extremes of the upper zones and perhaps, to some extent, on the nature of the deposit. The ragworm *(Nereis diversicolor)* and the amphipod *Corophium volutator* (*Fig. 12:1)* are common in all zones but the lugworm (*Arenicola marina), Scrobicularia plana, Eteone longa, Pygospio elegans* and *Peloscolex benedeni* do not extend into the upper zone. The cockle *(Cardium edule)* and *Capitella capitata* are confined to the lower half of the marsh. *Scoloplos armiger, Macoma balthica* and *Mya arenaria,* all common on the mud flats, are absent from the marsh. Two burrowing species are more abundant on the marsh than on the mud flats, one is the dipteran *Symplecta stictica* whose larvae are typical of salt marsh muds and soils, the other is *Tubifex costatus* an oligochaete worm which reaches its highest densities in the pans of the upper zone where the salinity is lowest.

Species living on or in the surface of the substratum include most common members of the mudflat epi-fauna e.g. *Hydrobia (Fig. 12:2), Littorina* spp., *Corophium (Fig. 12:1)* and *Carcinus (Fig. 12:3)* as well as others characteristic of more estuarine conditions which need more or less continuous immersion, for example, the small, filter feeding sabellid worms, *Manyunkia aestuarina* and *Fabricia sabella* both of which are sometimes very abundant. The rough winkle *(Littorina saxatilis, Fig. 12:4)* is represented on all parts of the marsh by black, brown, banded, tessellated and red forms, but the black type predominates.

Free-swimming fauna

The free-swimming fauna is generally typical of brackish waters. The small goby *(Pomatoschistus microps, Fig. 12:5)*, young grey mullet *(Mugil chelo)*, elvers *(Anguilla anguilla)* as well as shrimps *(Crangon vulgaris, Fig. 12:6)*, prawns (*Palaemonetes varians)* and the amphipod *Gammarus zaddachi* are all common estuarine species. Midge larvae *(Halocladius variabilis)* and copepods *(Mesochra lilljeborgi, Parathalestris intermedia)* may also be taken in the upper pools of a rocky shore. The free swimming fauna includes a small contingent of planktonic species which enter the channels with the rising tide and leave with the ebb. Most are small crustacea such as copepods and the isopod *Euridice pulchra;* their contribution to the food supply of fish may be important. Some species are particularly associated with green algae, especially herbivorous types such as midge larvae. Many species of the mud surface e.g. *Hydrobia, Littorina* spp., sea slugs *(Limapontia nigra* and *Alderia modesta), Peloscolex, Eteone,* sabellids and young nereids also congregate in algal masses.

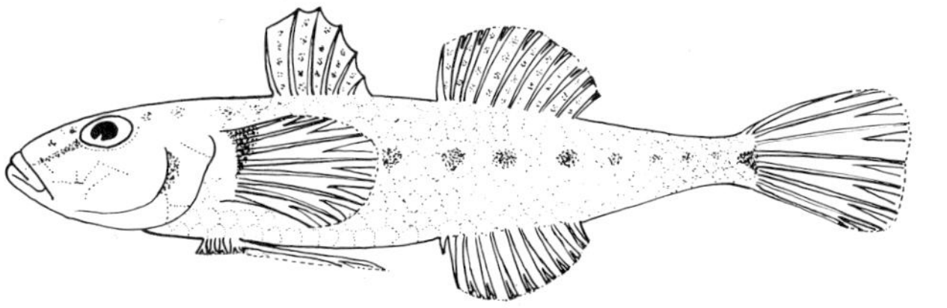

Fig. 12:5 *Pomatoschistus microps,* the small goby. (Natural size 3- 6·5 cm.)
Fig. 12:6 *Crangon vulgaris,* the common shrimp. (Natural size 2-7 cm.)

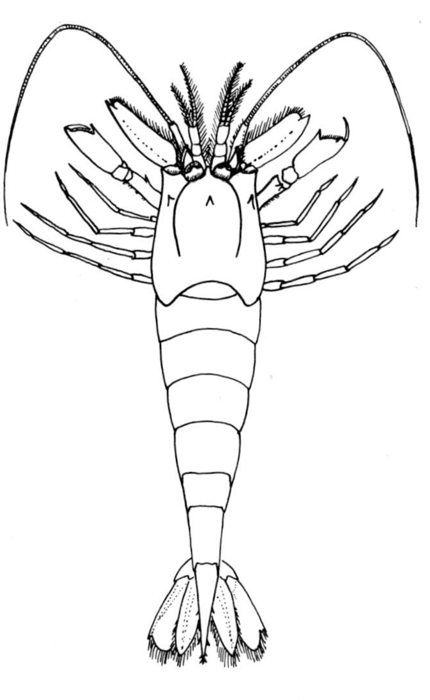

XV. The fern *Ophioglossum vulgatum* in the alder marsh with *Hydrocotyle vulgaris, Juncus maritimus* and *Poa pratensis.* (P. H. Pitkin)

XVI. Alder marsh in May showing a clump of marsh helleborine, *Epipactis palustris* beside a willow bush – *Salix fragilis* and surrounded by spikes of red fescue, *Festuca rubra.* (P. H. Pitkin)

The most conspicuous inhabitants of the channels are the shrimps and gobies, both of which, however, are to some extent migratory. Shrimps are winter spawners which come onto the marsh in summer. The life history of the goby on the Bull Island has been studied by Kunz (1969) who observed a seaward migration in spring of mature fish and a recolonisation in autumn by juveniles, movements being related to changes in temperature and salinity. Not all fish remain at sea during the summer, however, and it is possible to see shoals at almost any time of year.

Gobies are frequently seen to have external parasites on the fins or gill covers. These are the praniza larvae of *Paragnathia formica* an isopod with an interesting life cycle. The larvae remain attached to their hosts for about six months, then drop off and find their way to galleries in the banks of channels or in the cliff at the lower edge of the marsh. These have been dug by males and have entrances above HW ending in small chambers. Here the larvae remain for a while in colonies and then undergo metamorphosis. The females die after giving birth to larvae, while the males may leave and dig other galleries.

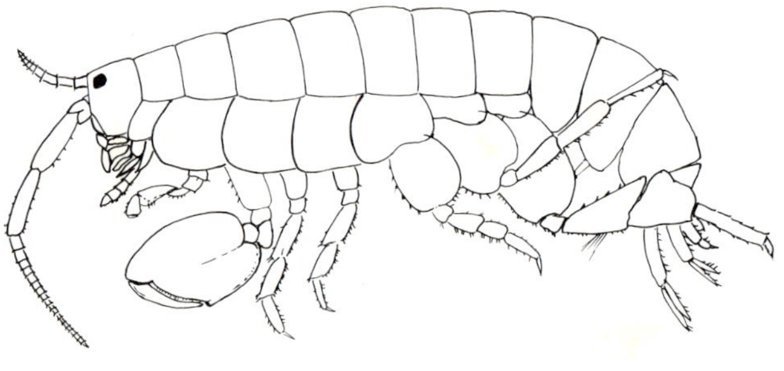

Fig. 12:7 *Orchestia gammarella,* a hopping amphipod. (Natural size 5-18 mm.)

Paragnathia is not the only species to find the banks a convenient and safe place to live. The banks of both pans and channels are in fact riddled with holes and burrows whose occupants remain in a moist microclimate during dry periods and find protection from birds. At the upper levels of banks, among the roots of plants, are seen the circular entrances, about 1 cm. across, made by the terrestrial amphipod *Orchestia gammarella (Fig. 12:7).* Below these are larger holes, often oval, made by juveniles of the shore crab (adults are rare on the marsh), which lead into a system of branched burrows. The excavations of crabs are often so extensive as to cause the banks to collapse. At a lower level again, near low water mark, the small holes made by *Corophium* are often abundant. The rough winkle is also found in banks but its burrowing abilities are limited and it tends to use crevices or crab burrows.

A characteristic feature of the aquatic fauna is the fluctuation in numbers of many species. The seasonal appearance of migrating shrimps and gobies has already been mentioned and other less common species such as prawns, eels and juveniles of the grey mullet also appear at more or less the same time each year. For other species, however, appearances are erratic and unpredictable. Large numbers may be present one day and practically none a week later. Two groups which are particularly subject to population fluctuations are the sea slugs and the sabellid worms. The probable explanation in most cases is that species may be well adapted to marsh conditions as long as these remain reasonably stable but are unable to withstand violent changes.

Finally, mention should be made of some terrestrial species which live in the vicinity of water. The best known of these is probably *Anurida maritima,* a small blue springtail which is common on most sea shores and is often blown onto water. It does not feed in water, however, and merely awaits a chance to return to land. Several marsh insects on the other hand actively attack aquatic animals. Predatory flies of the family Dolichopodidae are often seen resting on wet mud or even on the water surface where they devour emerging midges or insects trapped in the surface film. At least one common species, *Hydrophorus oceanus,* is known to "fish" victims out of the water. A group of predators with a different approach are the Saldidae or jumping bugs *(Fig. 12:10)* which run about on the mud at low tide and feed on small stranded animals.

Fauna of the pans

So far we have discussed the relatively stable fauna of channels. The fauna of pans is largely an impoverished version of that found in channels. *Nereis, Corophium, Hydrobia, Carcinus* and the two species of *Littorina,* all species which are able to

XVII. Assemblage of snails, *Cochlichella virgata,* on the deep rooted dune composite *Hypochaeris radicata.* Much vegetation has died back in the drought of 1975. (B. Healy)

XV XVI P. Pitkin

XVII B. Healy

XVIII D. W. Jeffrey

XIX P. Pitkin

XVIII. Dune slack grassland in May. The flowers or leaves of three nitrogen fixing legumes, *Trifolium pratense, Anthyllis vulneraria* and *Lotus corniculatus* are visible. The plant hemi parasite *Rhinanthus minor,* and the flesh coloured orchid *Dactylorhiza incarnata* are also prominent.

(D. W. Jeffrey)

withstand long periods of exposure, are present in nearly all pans. *Arenicola* and bivalves however are absent and populations of shrimps and gobies fluctuate. Some pans on the upper zone, near the tip of the island, are reached only by high spring tides and are mainly kept filled by rainwater. In the deeper ones, nereids, crabs and the oligochaete *Tubifex costatus* are the last marine species to survive. In the shallower pans where the water may be nearly fresh, even these disappear and small brackish water beetles are the only representatives of an aquatic fauna.

Fauna of saltmarsh surfaces

It is convenient to consider the terrestrial fauna under five headings:

a) an aerial fauna consisting of winged insects;
b) fauna of the vegetation—insects and arachnids found on plants;
c) a ground fauna consisting of species which live on or just under the surface of the soil;
d) a soil fauna and
e) the fauna of tidal debris.

Aerial fauna

Aerial species are independent of tides and able to range freely over marsh and dunes. A high proportion of the species recorded are not typical of salt marshes but have strayed into the intertidal zone from the dunes or have been blown onto the marsh from the mainland. The majority were taken in the upper zone which, apart from its proximity to the dunes is more likely to be favoured than the lower zones because the taller vegetation provides shelter and footholds for resting insects.

Flower feeding species are not generally abundant but two species breed on the marsh and are characteristic. One is a yellow and black soldier fly, *Nemoteles uliginosus,* which swarms over rushes at the end of June, the other is the rush moth, *Coleophora maritimella,* a small, drab species which is common all over the marsh in June and July. Most of the typical marsh insects are either predatory or non-feeding in the adult stage. Among non-feeding species, the non-biting midges are the most noticeable because the males form dense swarms over the channels on calm days. The females lay their eggs in the water or wet soil. Two species of small crane fly, *Erioptera stictica* and *Dicranomyia sera* breed in mud and soil and are common as adults but are inconspicuous because they spend most of their time on the ground. The better known large crane flies or daddy-long-legs *(Tipula paludosa)* are often abundant in September but this species does not breed on the marsh.

On rocky and sandy shores, seaweed flies (Coelopidae) are usually the dominant insect predators but these are rare on the marsh owing, probably, to the scarcity of brown seaweeds. They are replaced chiefly by the long-legged flies, Dolichopodidae, recognised by their greenish, metallic bodies, which are especially abundant on the taller plants and near water. The fishing activities of one species have already been noted. Another common predatory fly is *Scatophaga littorea* a close relative of the ubiquitous yellow dung fly which also frequents the marsh, may be seen from April to October, sitting on damp mud and on green algae in which it breeds.

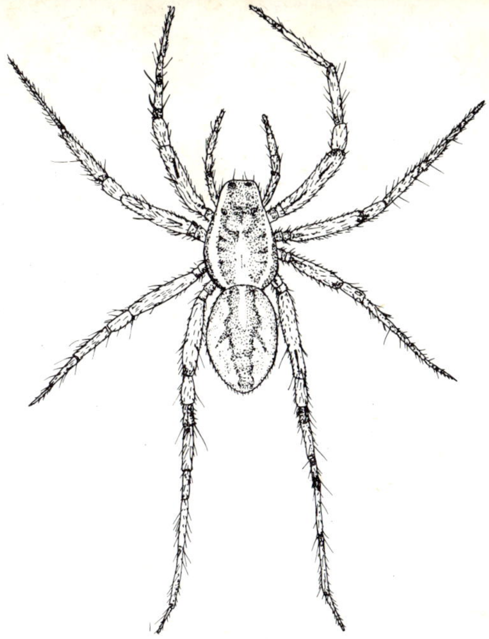

Fig. 12:8 *Lycosa purbeckensis,* a hunting spider. (Natural size 6-9 mm.)

Fig. 12:9 *Saldula palustris,,* a jumping bug. (Natural size 5 mm.)

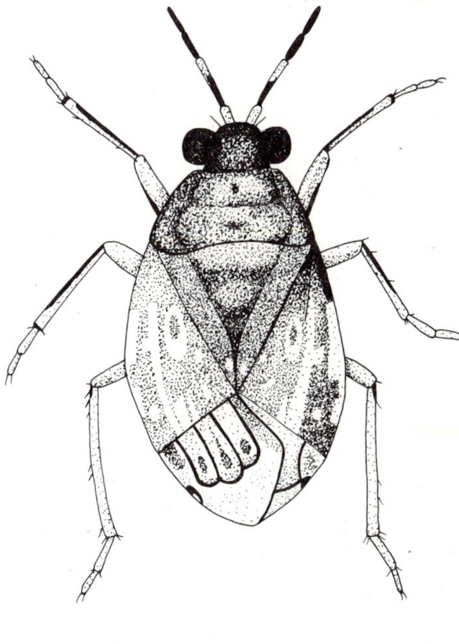

XIX. Marsh helleborine flower with a coastal species of cantharid beetle. This carnivore feeds on other flower visiting species.

(P. H. Pitkin)

The aerial fauna is distinctly seasonal, many species having a short adult phase lasting as little as two weeks. The first insects to be seen are usually *Scatophaga littorea* and the flower feeding March flies (Bibionidae) whose appearance coincides with the flowering of *Cochlearia officinalis*. The number of species then increases to reach a peak in July and August after which there is a falling off until the last individuals, usually *Scatophaga* disappear around the end of October.

Fauna on vegetation

The fauna of the vegetation, comprised of plant feeding species and their predators, does not become significant until the end of June when new plant growth is well developed. It is likely that many species, especially spiders and bugs, overwinter in the dunes because they first appear on the marsh in the upper zone and only gradually extend to the lower zones. The only species to be found all the year round are *Haplothrips statices,* a bright red thrips which lives on *Armeria* plants, hiding among the leaf bases in winter and climbing into the inflorescences in summer, and the larva of the rush moth which lives on the inflorescences of *Juncus maritimus* and is often so numerous as to constitute an infestation. The moth lays its eggs in the young seeds and the larvae, having eaten the contents of the seed, spin a protective cocoon around themselves and continue to feed on other seeds. They pupate within the cocoons, still in the inflorescences, and emerge in June. Plant feeding species which only appear in summer include other species of thrips, aphids (many of which are specific to certain salt marsh plants), froghoppers and plant bugs. Preying on these, in addition to predatory flies, are spiders, carnivorous bugs and soldier (or soft-wing) beetles.

Ground fauna

The ground fauna, although to some extent seasonal, comprises a high proportion of species which are visible throughout the year. Some of these are marine species which survive wherever the ground is sufficiently wet. *Littorina saxatilis (Fig. 12:4),* although most abundant in the channels, can actually breathe, eat and reproduce out of water and is found among plant roots wherever the soil is permanently moist. *Uteriporus vulgaris* an orange planarian is common in similar situations. In particularly wet patches, *Hydrobia* and even crabs may be present. A small, brackish water beetle, *Ochthebius punctatus,* which also occurs in shallow pans in the upper zone, is very abundant in damp depressions of the upper marsh where both adults and larvae burrow under algal mats. A number of mites are also semi-aquatic e.g. *Hygroribates bilineatus* and *Ameronothrus nigrofemoratus.* Only one land (pulmonate) snail occurs on the marsh, the estuarine *Phytia myosotis,* which is common in all zones. Ground-dwelling spiders and insects only become abundant between June and September. The most conspicuous are the spiders and predatory bugs. Two spiders are common all over the marsh, each with its distinctive hunting method. The estuarine wolf spider, *Lycosa purbeckensis (Fig. 12:8),* chases its prey and in calm, dry weather may be seen running about on dry pans and the banks of channels, the females carrying egg cases behind them. The smaller *Erigone longipalpis* spins small webs on the ground and is the only spider to remain on the marsh throughout the winter. The jumping bugs (Saldidae, *(Fig. 12:9),* of which several species may be distinguished, have been mentioned as feeding on the banks of channels but they too may also be found wherever there is bare, damp soil or mud. Less conspicuous, but equally important members of the ground fauna are, insect larvae. Many types may be recognised but most can only be identified by breeding through to the adult stage. Particularly common and characteristic are the maggot-like larvae of the small crane flies, *Erioptera stictica* and *Dicranomyia sera* and the flat, spiky larvae of the soldier fly, *Nemoteles uliginosus.*

Pitfall trapping has shown that many dune invertebrates extend their foraging activities onto the marsh at night, occasionally venturing as far as the middle zone. Three of the common species of woodlice *(Armadillidium vulgare, Philoscia muscorum* and *Eluma purpurascens)* have all been taken in large numbers and the carrion beetles *Silpha tristis* and *Thanatus rugosus* and their larvae are particularly abundant.

In view of the fact that the majority of ground dwelling invertebrates are typical terrestrial animals without obvious adaptations to aquatic life, one naturally wonders how these survive tidal immersion. It must first be said that many forms such as spiders (except *Erigone*) and bugs are only present in summer when tidal flooding is much less frequent than in winter and when surface water drains away quickly. The species which do survive frequent flooding use a variety of methods. Small light creatures with water repellant bodies such as mites and springtails are often seen floating on the surface during rising tides. One tiny oribatid mite, *Punctoribates quadrivertex* discovered by the Dublin naturalist J. N. Halbert, is particularly adept at floating and may travel some distance in this manner. Other species may seek out air pockets or simply assume a state of torpor until the tide recedes. Experiments with a variety of animals from the Bull Island and other marshes have shown that common species such as *Phytia myosotis,* saldids, mites and dipteran larvae can survive at least 24 hours immersion without ill effects. Indeed some mites e.g. *Hygroribates spoofi* and *Ameronothrus nigrofemoratus* continue to move about while submerged.

Soil fauna

If tidal flooding makes life on the ground hazardous, its effects on the soil fauna can be even greater. The soil becomes flooded not only from the surface by tides and sometimes by heavy rain, but also from below by the rising water table. However, percolation of water into salt marsh soils is so slow that the top few cms generally remain aerated throughout the tide cycle (Chapter 13), and constitute a refuge for air breathing invertebrates. There are few permanent members of the soil fauna in the lower zones of the marsh but whether this is due to frequent flooding or to high salt concentrations is not known. Only 23 species can be described as common, comprising 1 amphipod, 2 beetles, 4 enchytraeid worms, 4 springtails and 12 mites. The amphipod, *Orchestia gammarella (Fig. 12:7),* lives in vertical burrows at about 10 cm from the surface from which it emerges during tidal flooding and probably also at night. The two beetles are of particular interest. *Notaris bimaculatus* is a weevil whose adults and larvae feed on the roots of *Spartina* and which, in laboratory cultures, were never seen to emerge from the soil. *Dichirotrichus gustavii* in contrast, is a carabid or ground beetle and is probably carnivorous. Adults and larvae are common down to the sand layer in all zones. At night, adults emerge and climb onto plants, presumably to prey on other insects. In the *Festuca* facies of the upper zone where the soil is sandy and relatively dry, the soil fauna resembles that of the sand dunes with earwigs, wireworms and millepedes locally abundant.

Fauna of debris

A special habitat is provided by the piles of debris which are carried onto the marsh by tides. This is of two types: a 'dry' refuse consisting of the remains of salt marsh plants and flotsam from the sea which is mainly found in the upper zone, and *Enteromorpha* 'ropes' which are common in the lower zones. Dry refuse harbours a community of mites, springtails and small beetles, particularly Staphilinidae but the habitat is not very stable and is easily dispersed by winds. The *Enteromorpha* beds on the other hand remain moist for several weeks providing a nutritious medium which supports a rich fauna of dipteran larvae, beetles etc. The most important inhabitants are the maggot larvae of *Scatophaga littorea.* The life cycle of this fly, from egg to adult, takes six weeks or less and can thus be completed before the algal mass dries out.

The terrestrial component of the marsh fauna shares many species with the dunes especially near the limits of tidal penetration. Some dune species, however, have never been found on the marsh. None of the common dune snails has been found alive within reach of the tides and earthworms are entirely absent from marsh soils.

Acknowledgement

I wish to thank Mairéad Noone for Figs. 12:1–9

Chapter 13

ECOPHYSIOLOGY OF SALTMARSH PLANTS

K. A. McNamee and D. W. Jeffrey *Department of Botany* *Trinity College*

Introduction

To someone walking virtually dryshod on the saltmarsh for the first time in early summer, the firm texture of the middle and lower zones belies the term 'marsh'. Perhaps only the upper zone with conspicuous clumps of *Juncus* and soils spongy with much organic matter deserves description as such. The soils of all zones, in fact, normally have a water content of 60% – 80% and the firm texture of the lower and middle marshes reflects the sandy-silty composition of the soil and relative absence of much organic matter.

Two simple features of saltmarsh vegetation immediately indicate its unusual character.

(i) The 'crunchy' feel of the middle and lower marshes under foot is caused by the crushing of the succulent leaves and stems of plants like *Plantago maritima, Salicornia* and *Triglochin maritima (Plates II and XIII).*

(ii) The lowly habit of most marsh plants conceals an unusually high root : shoot ratio, often greater than 6:1 *(Fig. 13:1).*

The complex nature of saltmarsh plant associations, and the intriguing pattern of the soil environment have led to studies on the relationship between the two. This has entailed detailed observation of the variation of a whole range of factors, combined with experiments on the response of plants to them. Only in the laboratory can we separate factors which operate together in the field. The field does, however, provide an essential touch-stone for the verification of the relevance of experimental findings.

Factors affecting salinity

Tides and rainfall

The interstitial water in the marsh, particularly in the lower and middle marshes, contains high concentrations of salts, especially sodium. One of the effects of sodium in a soil is that it prevents aggregation of soil particles necessary for the development of a well structured soil. This, together with the silty-sandy soil composition, impedes drainage and slows down the penetration of tidal and rain water. Thus, tidal water may influence only the top few cm of soil. Such salinity, however, is not readily diluted or leached from the profile by subsequent rainfall. In an experiment which simulated tidal cover and rainfall we found for example, that a three-hour tidal cover of both the middle and lower marshes elevated interstitial salinity to the level of sea water in the upper 5 cm of soil only. Cover of the upper marsh for the same period gave rise to but 68% of sea water salinity in the same horizon.

In order to simulate the 30-year average April rainfall for Clontarf, 45 mm of distilled water was then applied to the soils in the field. This process took 20 hours in the middle marsh and about 2 hours in the upper marsh. The results shown in *Table 13:1* indicate the interaction between rainfall and salinity in the interstitial water.

Table 13:1, The effects of experimental seawater flooding and rainwater leaching on saltmarsh salinity. Results expressed as a percentage of seawater.

Site	Depth	Original salinity	After 3h seawater	After leaching	% Decrease
Middle Marsh	0–5 cm	94	100	72	28
	5–10 cm	88	88	80	9
Upper Marsh	0–5 cm	62	68	41	40
	5–10 cm	35	47	39	17

Evaporation and transpiration

An explanation of the varying salinity of a saltmarsh would not be complete without reference to the effects of evaporation from the soil surface and transpiration of plants. Both are influenced by the salt content of the interstitial water, the temperature (sunshine), relative humidity and wind. The rather wet summer of 1974 and the unusually dry summer of 1975, offered an excellent opportunity for comparison of salinity fluctuation in each zone.

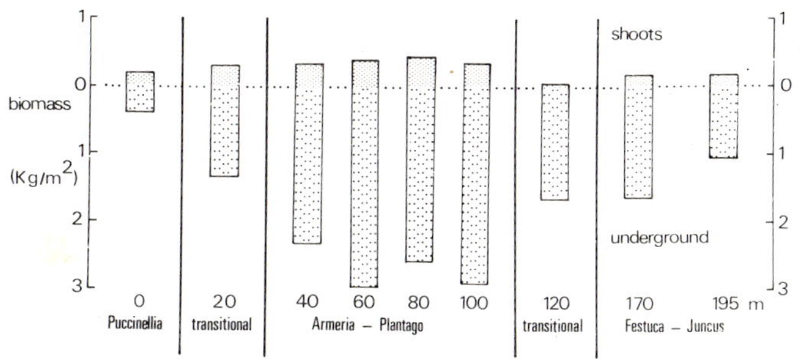

Fig. 13:1. Shoot and underground plant biomass for vegetation of salt-marsh zones. The high underground biomass is characteristic of vegetation from either arid zones or strongly saline terrestrial soils.

1975 Month	Lower Marsh	Middle Marsh	Upper Marsh
January	4, 8	9, 11	28
February	7, 8	22	26
March	8, 8	23	27, 2*
April	9, 8	23	30*
May	10, 5	25	31*
June	10, 5	14, 9	30*
July	9, 5	12, 7	31*
August	9, 5	11, 7	9,* 23
September	8, 5	10, 7	24
October	6, 4, 4	8, 8, 7	24
November	6, 2	7, 6	24
December	9, 3	11, 6	31

Table 13:2. Number of consecutive days of complete exposure between tides for each zone on the saltmarsh.
*Indicates continuous exposure.

Table 13:2 shows the number of days of continuous exposure of each zone between tides for the year 1975. The exposure period was calculated from the tables of predicted tide heights for the North Wall, Dublin, and confirmed by observation and records of actual tide heights supplied by the Dublin Port & Docks Board. In each case the number given refers to the length of time when the whole of the zone was exposed and would be higher if partial exposure were taken into account. The lower marsh was never exposed for more than 10 days at a time while the whole of the middle marsh could be exposed for up to 25 days. The upper marsh remained free of tide water for 4 – 5 months at a time.

Fig. 13:2 shows the very important influence of evaporation and transpiration during these periods of exposure. The water potential fell to nearly −50 bars in the middle marsh during the summer of 1975, a salinity twice that of seawater. Any tendency for the lower marsh salinity to increase by this amount was curtailed by the more frequent tidal cover. In the upper marsh the observed fluctuation was from almost fresh water to nearly full sea water, surely an influence of non-saline ground water.

Plants and their environment

It is possibly surprising to learn that some saltmarsh plants grow better in ordinary non-saline soil. Others, unlike most plants, require slightly elevated sodium levels for healthy vigorous growth. On the whole, however, salinities encountered in saltmarshes cause depressed growth even of saltmarsh plants.

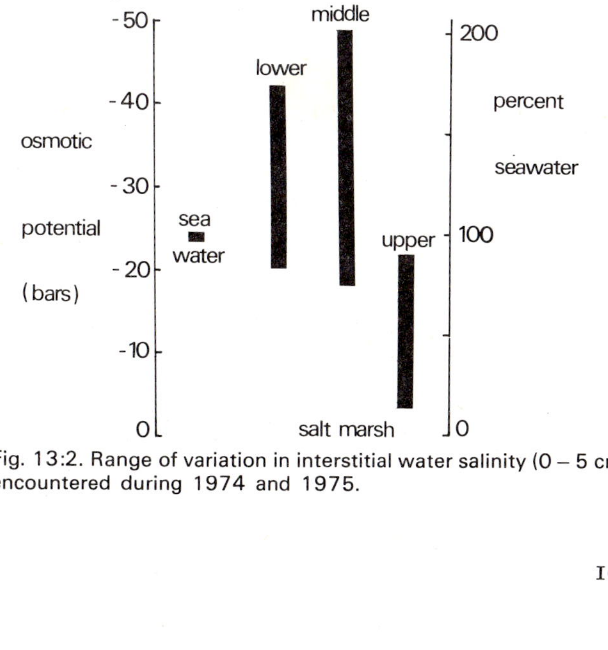

Fig. 13:2. Range of variation in interstitial water salinity (0 – 5 cm) encountered during 1974 and 1975.

Germination and salinity

In colonising saline soils, plants must not only germinate in these unfavourable conditions but must also establish themselves and survive to seeding stage. Their ability to survive in such conditions is a measure of their tolerance. It has been shown experimentally that germination of seeds of salt-marsh plants takes place best in fresh water or water of low salinity. *Table 13:3* shows the germination of a selected set of seeds from the saltmarsh, expressed as a percentage of the number which germinated in distilled water. It is interesting to note that the two upper marsh plants studied, i.e. *Juncus gerardii* and *Juncus maritimus* are more sensitive to sea water during germination than the four from the middle marsh.

Table 13:3	Percentage sea water concentration				
	20	40	60	80	100
Plantago	64%	54%	20%	1%	0
Triglochin	84%	37%	23%	15%	10%
Armeria	94%	82%	8%	1%	0
Halimione	104%	82%	37%	2%	2%
Juncus maritimus	99%	57%	8%	0	0
Juncus gerardii	38%	0	0	0	0

Table 13:3. The effects of seawater concentration on the germination of middle and upper marsh species from the North Bull.

Growth of Mature Plants

The effect of sea water on mature plants is two-fold. Firstly, adding ions to water lowers its osmotic potential (often referred to as 'raising its osmotic pressure') and makes it less available to the plant. Together with matric potential *(see Fig. 3:6)* it accounts for the limiting or restricting of the water supply to the plants. While matric potential is mainly responsible for water retention in dune soils, osmotic potential is the chief contributing factor in saturated soils of the saltmarsh. Their combined forces may be measured as water potential.

Secondly, the high concentration of a monovalent ion like sodium may interfere with the uptake of potassium and calcium and when absorbed into the plant may interfere with metabolism.

Fig. 13:3. The effects of salinity on the shoots of *Plantago maritima.* (a) and (b) show the relative sizes of young plants experimentally grown in full sea water (a) and one third sea water (b); (c) and (d) are cross sections through the leaves and show the corresponding structural modifications. Plants growing in full sea water (c) have short almost cylindrical leaves with a small number of vascular bundles while leaves of plants grown at low salinity (d) are dorsiventral with a large number of vascular bundles. These responses are paralleled by mature plants in middle and upper salt-marsh zones.

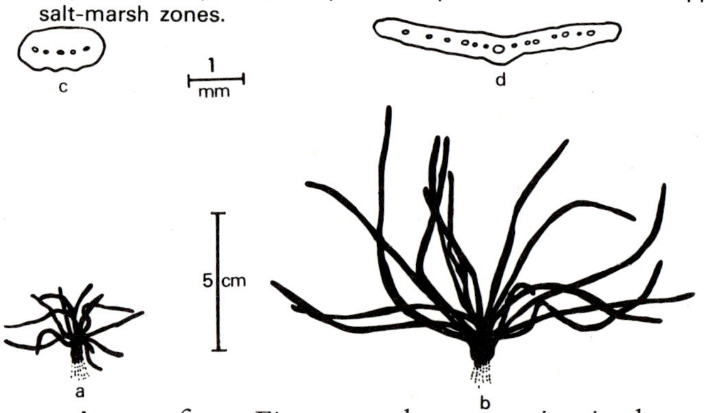

As seen from *Fig. 13:2*, plants growing in the saltmarsh, even the upper saltmarsh, have to adapt to very low water potentials. To obtain water in such conditions plants must change the osmotic potential of their cells to a more negative value than that of the surrounding water, the soil solution.

Middle and lower marsh species absorb large quantities of the readily available sodium and chloride ions which are then used to decrease the osmotic potential of their protoplasm. They are thus using, in a positive manner, ions which are known to be toxic to many plants.

All saltmarsh plants suffer a reduction in photosynthesis under conditions of water stress. As stress builds up some middle marsh species, e.g. *Plantago*, show a plasticity of response *(Fig. 13:3)*. Outer leaves are shed and the plant replaces them with shorter often more cylindrical leaves. These new succulent leaves have a smaller surface area, relative to volume, and therefore a smaller evapotranspiration area, so presumably cutting down on water loss through the shoots. At the same time they continue to develop their root system, increasing the absorption areas. Further explanation of the high root-shoot ratios seen generally in *Fig. 13:1* may lie in the provision of storage facilities for rapid replacement of non-succulent leaves.

Upper marsh plants respond osmotically to water stress but the reduction in length and number of leaves appears to be the result of reduced photosynthesis rather than a positive adaptation to saltmarsh conditions. If the water potential of the root medium which has decreased to the level of sea water is combined with temperatures of 20–25°C the shoots of these plants dry out and die.
In the field situation, many of the saltmarsh species, because of their deep roots, are able to make use of the more readily available water at a depth of

20 – 30 cm near the fresh water table. Sea water around the shoots of plants during tidal cover and even salt spray can cause leaf burn and death of non-halophytes. Some of the upper marsh plant species, e.g. *Festuca rubra* and *Juncus gerardii,* do show slight sensitivity to shoot immersion in sea water. In both, premature senescence of leaves occurs, with a consequent decrease in photosynthetic capacity.

Middle and lower marsh plants do not show this sensitivity and appear to be unaffected by tidal cover. Under conditions of high salinity, *Armeria* and *Limonium* secrete salt, and salt crystals may be seen on the leaves of *Armeria* growing in two-thirds or full sea water. These are re-dissolved during tidal immersion or by rain and dew.

Water-logging and reducing conditions (see ch. 3)

Water-logging and anaerobic conditions bring quantities of iron and manganese into solution that may be sufficient to kill sensitive plants. All parts of the marsh are not equally affected and the degree of oxygen depletion will depend mainly on the time of year and the porosity, organic matter and water content of the soil. Many saltmarsh plants have overcome this problem by developing aerenchyma which enables them to get, not only enough oxygen to their roots for respiration, but also enough to oxidise the surrounding soil, rendering toxic ions insoluble. Sheaths of light brown oxidised iron can be seen along roots in many parts of the marsh.

Plant species which do not have aerenchyma, e.g. *Festuca,* and *Halimione* can tolerate high concentrations of iron around their roots for limited periods, but when exposed to water-logging for long periods suffer from root damage and become yellow and stunted.

The nutrient status of salt-marsh soils

The supply of nutrients to vegetation from saltmarsh soils is relatively low. Increases in plant productivity are generally obtained if additional nitrogen and phosphorus is supplied experimentally. This implies that the deleterious factors to plant growth on saltmarshes, salinity and water logging, are not solely responsible for low levels of production. A current problem is deciding to what extent nutrient supply factors contribute to plant zonation. Several gross features of the saltmarsh soils are related to zonation from front to back of the marsh. Soil pH and carbonate content tends to decrease, a predictable effect of leaching also seen in the dune system. Total phosphorus is high at the front of the marsh and declines to the back, a situation either related to leaching or to the changed sedimentation pattern. Conversely, total nitrogen increases, a classic pattern in soil development but not readily explicable by the few observations made on nitrogen fixation. It is more likely that sedimentation of organic matter is a prime nitrogen input.

Whilst this represents the pattern of soil analysis, experiments on nutrient supply reveal a more complex relationship between soil and plant. A number of pieces of evidence indicate that more nitrogen is supplied to plants low in the marsh, *Salicornia* and *Sueada* for example, than elsewhere. Indeed, it is possible that restricted supply of nitrogen in more densely vegetated upper zones depresses their competitiveness. High nitrogen requirement and salt tolerance may even be linked in some fashion as yet unexplained. There is also evidence that the different total-phosphorus concentrations do not on the whole contribute to zonal vegetation differences.

Summing up

We are at too early a stage in getting to grips with the direct causes of plant zonation for dogmatic statements to be made. At the present time the main environmental elements seem to be salinity, soil aeration and nitrogen supply. The competitiveness, growth form and reproductive behaviour of plants, represents the response of the vegetation. The overall interaction, which is clearly complex, generates the zonal pattern. We can only repeat the comment made elsewhere in this volume 'much work remains to be done'.

Acknowledgements

Work on the water potential of saltmarsh species and saltmarsh soils owes much to the interest shown by Professor H. Wiebe, Utah State University. Some information is drawn from the moderatorship theses of Michael Fenner and Crenia Tooher.

Chapter 14
VEGETATION OF THE DUNE COMPLEX

J. J. Moore *Department of Botany* *University College*

It is quite evident that the seaward part of the island is made up of a series of parallel dune ridges, *(Fig. 2:1)*, with the youngest ridge towards the sea. A study of the plant communities as we pass from the strand to the grassland in the middle of the island enables us to see how important the plants are in stabilising this inhospitable, mobile sand and thus building up a stable, if evolving, ecosystem in which other forms of life can find a home.

The strand is a very difficult habitat for plants to colonise and it contrasts strongly with the saltmarsh in having no higher plants at all growing on the area between high neap and high spring tides. It is only on the zone above the HWST mark that plants begin to appear.

The first plants to colonise the sand occur sporadically within a zone which is approximately 4 m wide. The equinoctial spring tides deposit piles of seaweed and other jetsam to form the upper strandline. These piles are rapidly covered with a thin layer of wind-blown sand and these small hummocks form an ideal spot for the germination in late April of annual species—*Salsola kali* (saltwort); *Cakile maritima* (sea rocket) and *Atriplex* species. *Salsola* with its cylindrical fleshy leaves with a spine at the tip is the most common of these on the North Bull.

As these species grow, they accumulate more sand around their base, this being brought about by their acting as a tiny wind break. Once the wind-speed drops below a critical value, the wind-borne sand grains drop to the ground. These annual plants, with their roots safely embedded in the rich mat of decomposing algae below the surface, thus start the build-up of small sand dunes. However, these embryonic dunes will be quickly dispersed by the winter storms after the annual plants have died in Autumn unless a more lasting protection is provided.

Two perennial plants with a network of creeping stems and roots just below the surface are ideally suited to this second stage of the build-up of the dunes. These are *Honkenya peploides* (not particularly common on the Bull) and especially *Agropyron junceiforme* (sea couchgrass) *(Plate III)* which is very conspicuous on the front of the first dune ridge. Closely related to the scutchgrass, the scourge of the gardener, *Agropyron* sends its fast growing rhizomes through the embryonic dune, thus preventing its erosion. As extra sand is deposited, *Agropyron* is capable of rapidly growing up through the sand to the light.

The front of the first dune ridge is usually well colonised by *Agropyron* and evidence for the seaward growth of the island is provided by the scattered hummock of *Agropyron* which have survived the motor traffic and lie about 30 m from the front dune ridge. If the part of the island north of the causeway were to be protected from motor traffic, there is no doubt that the island dunes would extend by this mechanism further into the bay and thus continue the growth that has been observed over the past fifty years.

Once the *Agropyron* dunes have built up to a sufficient height to prevent the roots of plants from coming into contact with the salt water of the high tides, two further grass species enter, *Elymus arenarius* (lyme grass) and *Ammophila arenaria* (marram grass). These are much larger grasses and very vigorous growers. *Elymus* has now become fairly common on the North Bull although it was first noted only in 1922 by Brunker & Stellfox. Doubt has been expressed as to whether it is a native plant, but it does fit neatly into the pattern observed in other north-west European dunes where it occurs in an intermediate position between *Agropyron* and *Ammophila*. It seems that its arrival is another example of the successive colonisation of this relatively new island.

The marram grass *Ammophila* is actually favoured by being covered by wind-blown sand; it grows vigorously up through the sand and sends out numerous new side roots which enable it to make best use of the very limited supplies of nutrients present in the sand *(Plate IV)*. In some parts of Ireland the *Ammophila* dunes can reach heights of up to 50 m. (e.g. at Malahide "Island") but on the Bull Island the dunes are seldom higher than 10 m. In the central portion of the Island and also near the Bull Wall, the growth of these *Ammophila* dunes is seriously hampered by heavy pressure from visitors in the summer.

As we pass over the first ridge of the *Ammophila* dune a number of new species appear—perennials like the red fescue *(Festuca rubra)* and a number of yellow flowered plants of the daisy family (Compositae) such as the ragwort *(Senecio jacobaea)* and catsear *(Hypochaeris radicata) (Plate XVII)*. The spurge, *Euphorbia paralias,* is particularly characteristic of this stage but it only becomes common near the tip of the Bull, away from human pressures.

There is relatively little accumulation of sand on the second dune ridge since most of the wind-blown sand is trapped by the first dune. A closely woven mat of mosses covers the second ridge and prevents sand blowing away. In early Spring this ridge is usually dotted by the tiny flowers of a number of ephemeral annual species like the tiny *Veronica agrestis* (Speedwell) and the smaller white flowers of *Erophila verna.* These plants avoid summer drought by completing their life cycle in a few weeks. Later, the sand-dune violet *(Viola tricolor* ssp. *curtisii)* whose petals are coloured in all possible combinations of yellow, blue and black, becomes obvious.

In this zone we note the first of the leguminous plants whose roots have conspicuous nodules inhabited by nitrogen-fixing bacteria *(See Chapter 3)*. The two most obvious of these species are the yellow kidney vetch *(Anthyllis vulneraria)* and bird's foot trefoil *(Lotus corniculatus) (Plate XVIII)* while the pink rest harrow, *Ononis repens,* is fairly common.

A thin surface layer of humus made up of the dead stem bases of the mosses and the decaying roots of the annual higher plants builds up rapidly on this ridge and prevents erosion of the sand. However, this skin of humus is quite thin and fragile and if it is destroyed by a fire or by too much trampling, the sand can be eroded thus leading to a 'blow-out'. This 'blow-out', or eroded hollow can be rapidly enlarged in windy, dry weather since the wind produces a vortex in the hollow that can rapidly scoop away the remaining sand.

As we move further towards the centre of the Island we reach dune ridges which are more than 20 years old and here we can see the development of a much more stable and lush vegetation while the soil will have a deeper layer of humus in which a dense mat of perennial grass roots guard against erosion. A number of grasses which can withstand the Summer drought dominate the scene—red fescue, quaking grass *(Briza media)* and meadow grass *(Poa pratensis)*. A few nitrogen-fixing clovers and vetches are also present. In early Summer the most striking flowering plants here are the yellow rattle *(Rhinanthus)* and eyebright *(Euphrasia),* which are semi-parasitic on the roots of the grasses; later a number of orchids appear, the most obvious of which is the Pyramidal orchid *(Anacamptis pyramidalis)*. If one is lucky one may see the bee orchid *(Ophrys apifera)* where the lower lip of the flower is a perfect imitation of a bumble bee sitting on the flower. This is an adaptation to pollination since male bees are misled and attempt to copulate with the supposed female and thus pick up the pollen from the flower and in their further attempts at copulating with other flowers transfer the pollen.

The two Golf Clubs have modified much of this dune grassland zone for their fairways; the addition of fertilizers and continuous mowing have produced a rather different combination of plants more characteristic of dry suburban lawns. However there are areas in the 'roughs' where a number of interesting species, especially of non-flowering plants, are known to occur and are unwittingly protected by the Golf Clubs.

About 1 km. from the tip of the Bull Island a large dune slack with a rich flora occurs, usually referred to as the 'Alder Marsh' because of the presence in it of about 200 Alder trees. This is an area of low-lying ground about 700 m. long × 70 m. wide situated between the dune ridges where the water table is very near the surface. This ground water is nearly fresh with but very slight salinity. A large part of the marsh is dominated by the saltmarsh rush *Juncus maritimus* but in other areas there is a terrestrial marsh vegetation characterised in Summer by meadow sweet *(Filipendula ulmaria)* and by a rich orchid flora including *Epipactis palustris (Plate XVI)* and *Dactylorhiza incarnata* of a delicate magenta hue. The little adderstongue fern *(Ophioglossum vulgatum)* may also be found *(Plate XV)*.

The distribution of all these different plant communities may be seen in the vegetation map *(Fig. 10:1 a & b)* and the more characteristic species of each type are summarised in Table 14:1.

Fore shore	Fore dunes	Primary dune ridge	Fixed dunes		Dune Slack
Cakiletea	*Agropyretum*	*Ammophiletum*	*Tortulo-Phleetum*	*Dune grassland*	

Cakile maritima
Salsola kali
Atriplex glabriuscula

Agropyron junceiforme

Ammophila arenaria
Elymus arenarius
Senecio vulgaris

Viola curtisii
Tortula ruraliformis
Phleum arenaria
Anthyllis vulneraria
Aira praecox
Cerastium diffusum
Trifolium arvense

Luzula campestris
Plantago lanceolata
Briza media
Trifolium pratense
Helictotrichon pubescens
Vicia angustifolia
Ononis repens
Polygala vulgaris

Festuca rubra; Poa pratensis;
Galium verum; Lotus corniculatus;
Rhinanthus minor; Senecio jacobea;
Hypochoeris radicata; Euphrasia occidentalis.

Juncus maritimus
Glaux maritimus
Hydrocotyle vulgaris
Cardamine pratensis
Agrostis stolonifera
Filipendula ulmaria

Table 14:1

Chapter 15

INVERTEBRATES OF THE DUNES AND GRASSLAND

Martin C. D. Speight *Department of Zoology Trinity College*

The sand-dune environment

To small animals sand-dunes represent a rather unfavourable environment for two main reasons: first, there is very little vegetation cover at the actual ground surface, second, the raw sand that constitutes the dune 'soil' contains little organic matter and does not hold water. In consequence, although the sand warms quickly, it just as rapidly cools, or dries out. These characteristics prevent many otherwise ubiquitous grassland animals from colonising dunes, while others are excluded at least from the marram grass areas because of the impoverished nature of the dune flora there. The Bull Island dune and dune grassland fauna is as subject to these influences as are dune faunas elsewhere. Thus there are two main elements in the Bull Island fauna: the residue of generally distributed grassland animals that can cope with dune conditions and the duneland specialists, a limited assemblage of species rarely found away from sand-dune or coastal locations.

Grassland animals surviving in dune environment

Nearly 50% of the animal species so far recorded from the dunes and dune grassland of Bull Island are widely distributed grassland species. The cuckoo-spit bugs *Philaenus spumarius* and *Neophilaenus lineatus* which feed on plant juices are good examples. The frothy blobs made on grass stems by their nymphs are found in the summer almost everywhere that taller grasses grow in Ireland, including Bull Island. Most of the plant-feeding species among these general grassland animals feed on a wide range of plants. The large black slug *Arion ater (Fig. 15:1)* is typical of this group. An exception is provided by one of the more conspicuous Bull Island insects, the red and black, day-flying, six-spot burnet moth *Zygaena filipendulae, (Fig. 15:2)* whose larvae feed on bird's-foot trefoil (though they can be found consuming certain other Legumes too). Both *Arion* and the burnet caterpillars are leaf chewers, while the cuckoo-spit bugs are juice-suckers. Another type of plant feeder is the seed-eater, exemplified in the grassy areas of the Bull by the common red ant *Myrmica ruginodis,* which plods around among the vegetation collecting grass seeds.

Generally distributed grassland detritivores are represented by the woodlice *Philoscia muscorum* and *Armadillidium vulgare,* which feed on plant remains and the micro-organisms found on decaying plant material. Predators are typified by the grassland hover-fly *Platycheirus clypeatus,* whose larvae crawl over the low-growing vegetation feeding voraciously on green-fly (aphids) that occur there, even on the marram grass. Other grassland animals, like the honey-coloured bumble-bee, *Bombus muscorum,* use Bull Island as much as a base as a feeding ground, since even though their nests may be hidden under the long grass there they can forage far away in Sutton, or Dollymount.

Most of the grassland animals occur more in the dune grassland areas of Bull Island, leaving the

Fig. 15:2 The six-spot burnet moth *Zygaena filipendulae,* its caterpillar and cocoon.

Fig. 15:1
The giant black slug *Arion ater.*

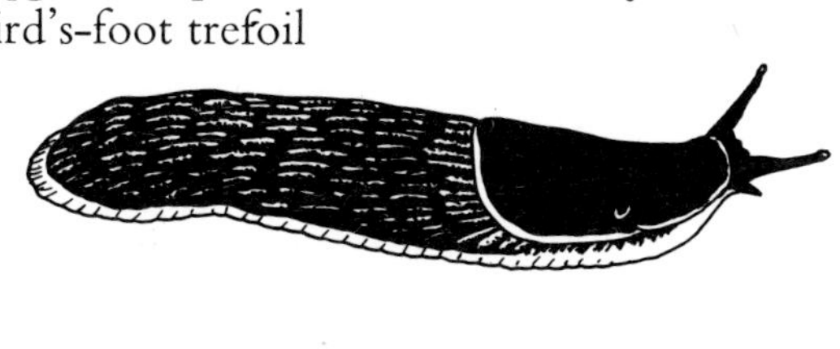

marram-covered dunes primarily to the duneland specialists. The ground-beetle *Laemostenus terricola* is a rather enterprising exception. Along the strand-line this beetle feeds on the insects that accumulate to feed on organic debris left by the tide, or the rubbish left by people. In grassland it is to be found in a variety of situations, such as the debris under wasps' nests, feeding on the remains of wasps' meals dropped down from above and on other scavengers present. In addition it has been recorded from the cellars of old houses! Other grassland animals which penetrate into the dunes proper are to be looked for in the slacks, such as the "alder marsh", where the thicker vegetation cover and damper ground surface provide conditions approximating more closely to those generally occurring in grassland. Here is to be found the grassland snail *Vallonia pulchella* and the tiny ground-beetle *Trechus obtusus,* the latter occurring all round Dublin in damp grassland. The gangling, elongate, grass-Mirid bug *Leptoterna dolabrata,* whose nymphs look like ears of grass seeds, is another characteristic damp grassland insect common in the Bull Island slacks. Both nymphs and adults of this species pierce grass blades with their syringe-like mouthparts and suck out the juices. Like many other bugs, *Leptoterna* overwinters in the egg stage, the eggs being tucked into the base of grass stems.

One of the commonest flies found in the alder marsh is a species of Sciomyzid, a snail-killing fly. This insect, *Pherbellia cinerella,* has larvae which actively seek out living snails and attack them, burrowing into their flesh to kill them. Each *Pherbellia* larva then feeds on the decaying tissues of its host for some days before moving off to attack another snail. The entire life-cycle takes only 30 days and because of the mild winters, at places like Bull Island these flies occur throughout the year, though elsewhere (they have a wide distribution, being known from Afghanistan to Greece and Norway to Ireland) they over-winter as pupae hidden in the dead snails. To what extent the adult flies found in the "alder marsh" breed there is a matter for speculation, since *P. cinerella* is also abundant in the saltmarsh and can be collected in smaller numbers over the whole area of Bull Island. In Ireland this is a widely distributed damp grassland insect, but it is particularly abundant along the coast.

Duneland specialists

Among the sand-dune specialists are the most local invertebrates so far recorded from the Bull Island grassland and dunes. These are both species of fly, the bee-fly *Phthiria pulicaria* and the hover-fly *Sphaerophoria rueppellii. P. pulicaria* is a tiny grey fly recorded only twice from Ireland, its second Irish locality being the dunes at Portmarnock. It is said to be found sitting on bare sand in hollows in the dunes. The seeming rarity of this little insect is probably more due to its insignificant appearance and the lack of Dipterists in Ireland, than to anything else! *S. rueppellii,* first recorded in Ireland during the course of this study of the Bull Island fauna and flora, is much less likely to have been simply overlooked previously, though scattered colonies probably exist elsewhere in E. coast dune systems. This elongate, black and yellow fly occurs locally in England and Wales, usually in coastal dunes, but nowhere there does it appear to be common. It is an animal of dune slacks and is found on Bull Island in the 'alder marsh'.

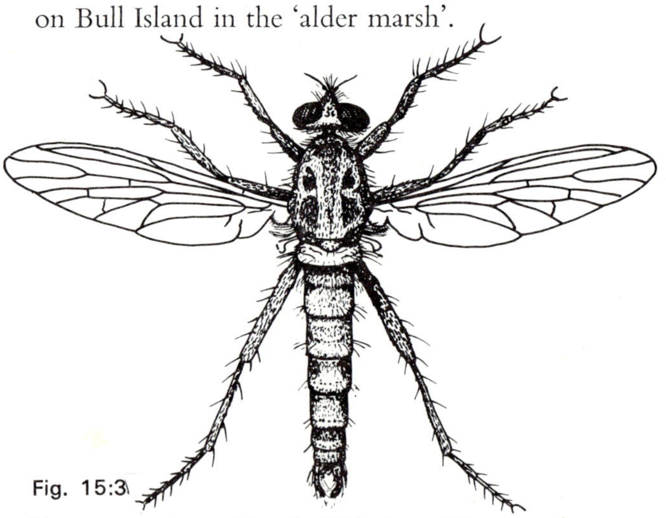

Fig. 15:3
The greater dune robber-fly *Philonicus albiceps.*

One of the largest of the sand-dune animals found on Bull is the conspicuous and rather noxious-looking, bristly, grey-brown robber-fly *Philonicus albiceps (Fig. 15:3).* This creature sits crouched on the bare sand on paths through the dunes, when approached whisking away only to settle again some feet further on. It is predatory and takes up station on paths to lie in wait for unsuspecting insects that fly over-head, at which point it makes a short capture-dart into the air, grabbing its prey in flight.

At close quarters its appearance is made all the more daunting by its long, horny proboscis, all too reminiscent of the mouthparts of a horse-fly, and used to plunge into its prey to suck out the body contents. Rather smaller than *Philonicus,* but found in the same situations are the three Irish species of *Thereva,* all of which occur on Bull Island. *T. nobilitata* is a most beautiful insect, being entirely covered in silvery-grey fur. It has the same habit as *Philonicus* when disturbed but is very rapid in flight, appearing to the eye as a silver flash which disappears when the fly lands, so closely is it camouflaged by the glaring white of the sand. One of the most busily pre-occupied of the dune animals is the spider-hunting wasp *Psammochares plumbeus,* to be seen running backwards and forwards over the bare sand with a peculiar jerking gait vibrating its antennae rapidly, for all the world like a dog trailing a scent (not that a dog has antennae!). This is precisely what the wasp is doing, following some luckless spider! The spiders are stung and then stored in burrows for the wasps' larvae to eat.

Fig. 15:4

The sand-hill snail *Cochlicella acuta.*

A rather different, but just as noticeable, denizen of the dunes is the sand-hill snail *Cochlicella acuta (Fig. 15:4),* whose shell is a narrow, pointed spire, whitish with brown bands across the whorls. On hot, dry days in the summer, clusters of these snails are to be found up among the heads of ragwort and other plants, having climbed there to take advantage of the slightly cooler breeze away from the hot ground surface. Another duneland snail with flattened, banded whorls, *Helicella virgata,* adopts the same strategy.

Grubbing around among the roots of marram grass reveals further elements of the fauna not normally encountered in any other way. Small Linyphiid spiders, like *Typhochrestus digitatus,* are common there. Particularly characteristic of this habitat are two little straw-coloured ground-beetles, *Demetrias atricapillus* and *Dromius melanocephalus,* which clamber back up among the marram shoots, for they live actually up in the vegetation rather than on the ground like most species in their family. Although not restricted to dune systems these are both very much coastal insects. One of the most bizarre inhabitants of the older, thicker, marram tussocks where there is a layer of dead marram grass is the little bristle-tail *Dilta hibernica,* a greyish, shining, carrot-shaped animal with what appear to be antennae at both ends. When disturbed *Dilta* flops and runs about till it has managed to hide again. Bristle-tails are representatives of a very ancient group of wingless insects. Little is known about the Irish species.

Many of the sand-dune animals are nocturnal, remaining hidden at the base of marram clumps or buried in the sand, during the day. This enables them to avoid the desiccating effects of the sun. They can be uncovered by searching inside marram tussocks during the day, but it is far easier to find them by torch-light, as they wander over the sand at night. This is the way to find the pale dune woodlouse *Armadillidium album.* One of the largest invertebrate predators inhabiting this zone roves the dunes at night: the black and aggressive ground-beetle *Broscus cephalotes.* This beetle will attack caterpillars, woodlice, other beetles, sand-hoppers and any other small animal that comes its way.

The dune slacks

Surprisingly few dune-slack animals have been recorded from the slacks on Bull Island. One of them, *Sphaerophoria rueppellii,* has already been mentioned. Another is the cuckoo-spit bug *Neophilaenus exclamationis,* which is found in the 'alder marsh' together with its more widely distributed relatives. This species is also an animal of salt-marsh, though it is not known from the Bull salt-marsh areas.

Animals with curious distributions

Some animals which in Ireland are restricted to coastal locations occur more widely elsewhere and certain of the dune grassland species present on Bull Island are good examples of this phenomenon. The hairy woodlouse *Eluma purpurescens* is an extreme case. It is known from cliffs and dunes in Co. Dublin but is not found further North than Portugal in the rest of Europe. Then there is the hover-fly *Sphaerophoria scripta,* recorded in Ireland only from scattered localities round the coast, but found throughout Southern England. And the minute snail *Pupilla muscorum* occurs in drier grassland of all sorts in England and Wales though not normally found away from the coast in Ireland. All three of these species are inhabitants of the fixed dune grassland on Bull Island, though *Eluma* is also present there in other habitats. In all probability many factors rather than one operate to restrict the distribution of animals such as these in Ireland, though it is tempting to speculate that lower rainfall, higher ground temperatures and drier ground surface are of prime importance, since these species have evidently a 'Southern' distribution in Europe. The restriction of *Eluma* to Co. Dublin could well have a different cause though: it is perhaps a species accidentally introduced to Ireland by man in the recent past.

Man-associated animals

Although the two main elements in the invertebrate fauna of the dunes and dune grassland of Bull Island are species characteristically associated with grassland and sand-dune habitats, some 20% of the recorded species do not really belong to either of these categories. These are nearly all synanthropic animals, i.e. ones particularly associated with man's habitations and their environs. They might conveniently be styled suburban species! One very well-known representative of this group is the garden snail, *Helix aspersa,* which with a trio of garden slugs *(Milax budapestensis, M. gagatea, M. sowerbyi),* occurs in the vicinity of the Corporation dump on Bull Island. The garden snail is now widespread over the island. The extent to which fully-flighted synanthropic insects owe their presence on Bull Island to the proximity of suburbia or to the Corporation dump is arguable, though at least one, the filth-feeding hover-fly *Syritta pipiens,* rarely occurs more than a few yards from its breeding sites in organic debris and to date has only been recorded from immediately by the dump. The opposite extreme is exemplified by another hover-fly *Episyrphus balteatus,* which although primarily synanthropic in the British Isles (its larvae are important predators of green-fly on suburban rose bushes!) is highly migratory and turns up just about everywhere throughout Eurasia.

Alder bush animals

One tiny group of species belongs to none of the categories mentioned so far: its members are dependent upon the alder bushes present in the 'alder marsh' and are thus more typical of damp woodland. The small, green, jumping bug *Psylla foersteri* lives only on alder, and can be found in many different situations so long as its food-plant is present. The beautiful pearly-green lacewing *Chrysopa ventralis* is characteristic of scrub woodland, where it may be found on a variety of deciduous trees and bushes.

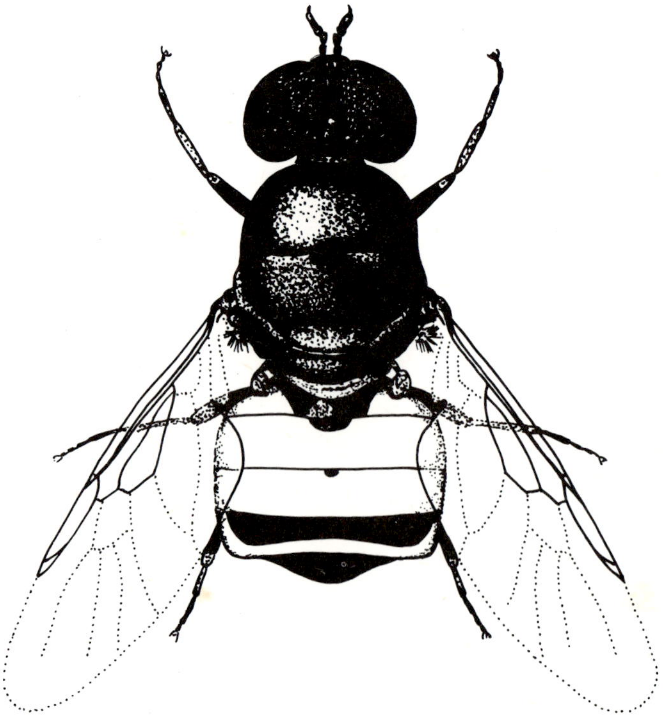

Fig. 15:5

The salt marsh soldier-fly *Nemotelus uliginosus.*

Non-residents

This account has concentrated upon animals that may be regarded as 'resident' in the grassland and dunes. Inevitably there are other species which may be found there that are 'just visiting' from other habitats on the island. This is particularly true of flower-feeding insects. For instance, the small soldier fly *Nemotelus uliginosus (Fig. 15:5)* which breeds in the salt-marsh occurs in numbers at flowers in the 'alder marsh'. Similarly, a most attractive little beetle, *Psilothrix cyaneus,* which is metallic green or blue, may be found in great abundance feeding at yellow-flowered Compositae throughout the dunes. Again, this species originates primarily from the salt-marsh grassland.

The state of our knowledge

The list of invertebrate animal species known from the dunes and grassland of Bull Island is at present woefully incomplete. Spider records are almost entirely lacking, the beetles have been recorded only haphazardly, the soil faunas are virtually unknown. So can anything more be deduced from what we *do* know? The animals that are recorded may be regarded as a sample of the total fauna, a sample selected without any particular end in view. That being so, there is no reason to suppose that this sample gives a distorted impression of the total grassland and dune fauna, in respect of its ecological characteristics and general interest. However, one important question cannot be given an unequivocal answer, namely, how depauperate *is* the fauna? Assuming the commoner animals of the dunes and grassland have been recorded, there are some surprising absentees, including rarities that are locally common in the dunes at Portmarnock and Portrane. One of these rarities is the robber-fly, *Epitryptus cowini.* Another is the snail *Theba pisana.* Among common species apparently absent is another snail *Cepea nemoralis,* recorded from right beside the coast road near the landward end of the causeway to Bull Island! A proportion of these puzzling absentees may well turn up on Bull Island once its fauna is further studied, others may be absent because of the existence there of inappropriate environmental conditions which we cannot perceive, but some of them could easily be absent due to the recent origin of Bull Island, i.e. they have yet failed to colonise the area.

Chapter 16
MAMMALS OF THE ISLAND

F. O'Gorman *Department of Pharmacology Trinity College*

As with most islands, the mammal fauna of North Bull Island is depleted in relation to the neighbouring mainland. Inevitably, some mammals just cannot or will not make a sea passage and, as a result, island populations of mammals tend to be of the smaller species, such as mice or rats which have come in accidentally with man, or others such as the rabbit which have been introduced deliberately. Of course with an island such as the Bull, which up to recently had no solid connection with the mainland, except for the wooden bridge, the mammals which arrived either came there inadvertently or came across the mud flats at low tide. At low spring tides, before the causeway was built, there was only a small channel of water probably not much more than three or four yards wide and six to nine inches deep, separating the mud flats on the mainland from the mud flats on the island.

There can be no doubt that this is the way the hares, which are the most visible feature of the mammal fauna of Bull Island, arrived. In fact, there have been a number of incidences recorded of hares seen crossing from the mainland. I have even seen a hare not only crossing the mud flats, but also crossing Bull Bridge.

The Irish hare, *Lepus timidus hibernicus* is the only sort of hare found in the Republic. There are no authentic recent records of the Brown Hares, *Lepus europaeus* anywhere in the Republic though there are a number of accounts of introductions in the past. In Britain and the rest of Europe *Lepus timidus* tends to be a mountain animal while the brown hare is found in the lowlands. In Ireland the Irish hare is found from the sea shore to the top of the Macgillicuddy Reeks in most habitats where it can graze, whether on grass or heather, or even among the intensly monocultured conifers of the Wicklow Mountains and elsewhere. However. it is rather thinly dispersed on the whole and that is what makes the Bull Island population pretty unique. Here you can see, if not hosts of hares, at least groups of 5, 10 or even 20 on occasions *(Plate XIV)* During spring, these mutually interacting groups are most obvious, though the largest populations are in the autumn at the time of the dispersal of the young. This is the time when one has the best chance of seeing an attempt to cross the mud flats to the mainland.

The Irish hare is called the blue or varying hare because this species tends to turn white or at least go bluish grey in the winter in the rest of its range. Our hares rarely turn white or even partially white except in the North or Northwest, but in Scotland, where there is generally a long period of snow cover in winter, the majority do. On the Bull Island the hares merely become duller and less russet. In their summer coat they are often a rich copper which only tends to confuse, since people seeing these rich brown animals often think they are the brown hare. But, as so often in these cases, the differences are relative—you really need the two species side by side to see the differences. However, the flat top to the ears and the white scut are the main characters to look for in the field. Incidentally you can distinguish a hare from a rabbit by the fact that hares' eyes always have yellow irises while rabbits have brown ones.

One of the pressures on the Bull Island hares is that from man. Continual disturbance by dog chasing—even illegal coursing with greyhounds has taken place—has resulted in their injury and death. They are quiet, delightful animals. One should approach them with caution and enjoy the pleasure of seeing one of the delightful scenes of Irish wildlife at close quarters and without disturbance.

One of the enigmas of the Bull is the rabbit. According to Eugene O'Mahony, who has written the only account of the Bull Island mammals in the last fifty years, rabbits used to be plentiful along 'Lady Anne Bank' and in the rocks of the breakwater. At the time he was writing (1931) the only place he ever saw one was 'about the rocks at the breakwater and then only early in the mornings'. He puts down their decline to ferreting.

As far as we know, the only other mammals on the island are the house mouse and the field mouse. Surprisingly enough, no rats or stoats have been recorded on the island. You would think that with a flourishing small mammal population, and plenty of cover the stoat would appear the obvious addition to the island's fauna.

In the past there was no solid connection with the mainland and the physical barriers for such a small animal like the stoat would have been insurmountable. The wooden bridge consists of an open lattice work of cross beams and together with the gelatinous mud flats would have defied anything but the most determined assault. However, neither need deter the fox, but it has not been recorded either.

Similarly, the lack of suitable habitats has prevented the establishment of any bat colonies, though no doubt bats do occasionally cross to the island.

Despite the spectacular nature of the Irish hares which now abound and whose populations have been variously calculated to be in the hundreds but probably are not more than 50–70, even in their best years, the island is far more famous for its mouse. Most Dubliners know the Bull Island as a beach resort, or a place for wintering wildfowl, but scientifically the area has worldwide renown because of the 'Bull Island Mouse'.

Because the island is of known age the maximum period the mice can have been living on it is also known. This is a rare occurence. Normally one has no idea how long a particular population has been in existence. During this period of say 160 years or so, they seem to have developed some unusual characteristics. These discoveries were made towards end of the last century by Mr. Lyster Jameson, not long after the storm about evolution between the Darwinists under Thomas Huxley and the Church under Bishop Wilberforce and, as a result, the fame of the Bull Island mouse spread far and wide. Here was a case of evolution taking place within a known period, in fact at that time, within only 70 or 80 years. The mice were very pale sandy coloured, really quite unlike the normal house mouse which is a rather greyish brown animal. The apparent explanation was that they were living in a sandy area and as a consequence had to adapt to that habitat to survive predation by hawks. So here was a straightforward case of evolution in practice. Such an example in a mammal was (and is) virtually unique. One of the few similar sorts of case is the one which relates to a moth in industrial Britain where melanism has spread through the population since the dirt and smog of the industrial revolution has blackened the countryside. Nowadays these moths are mostly black instead of being peppered as they would be normally. Here was a case of environmental changes causing a genetic response in this particular moth. While detailed studies of these moths has taken place no such studies have been directed at the Bull Island mouse.

Jameson described his findings in August 1889 and published a coloured painting of the animal which was later in 1929 to be named a separate species and called after him – *Mus jamesonii*. Efforts were made to collect more mice on the Bull Island in the nineteen thirties by O'Mahony but only a very few were caught, and the next effort was made by Azzo von Rezzori and myself in the early sixties. The work was prompted by the building of the causeway in 1966. I knew from my previous trapping on the mainland that the refuse from which the causeway was being made was infested with house mice at the mainland end. As the causeway advanced towards the island I could see this horde of mice poised all ready to pollute, in every sense of the word, the pure strain of Bull Island mice, which had existed there up to that time. So, desperate efforts were made to try and provide some baseline material which would then be available against which to check new collections in say, ten, fifteen, or twenty years' time, exactly in the same way as we had used the Jameson collection to check what has happened between 1890 and the mid sixties. It was to be a rush against the invader: were we going to collect enough animals to allow valid comparison to be made before they (and we) were overtaken by the advancing hordes?

We set out to see if variations which Jameson had shown in his collection had increased or remained stable in the preceding seventy years. Fortunately the collection that Jameson had made on the Bull Island was still safely in the National Museum. We were able to use new techniques to test this sort of 'epigenetic' variation as it's now called. Anyway, the upshot of all this was that we discovered really quite a significant difference in the colouration of the animals which now lived on the island and those which had been described by Jameson. Ours were much whiter underneath and browner on the back and not sandy like those of Jameson. So the cat was among the pigeons again. What had happened in between? Was this a progression? Was it something new? was 'genetic drift' just flowing to and fro in the population? What could have happened?

In the event we caught some thirty house mice, enough to allow some tentative conclusions. I now feel that Mr. Jameson's mouse was never really a valid species or even a subspecies or race but was, and is, the result of an introduction from the mainland of a small group of mice, perhaps only one or two, which due to inbreeding has caused this colour 'aberration' to spread throughout the population. The field mice, which are abundant, show no signs of being different from their mainland cousins. One can only presume the same origins for both species but that initially greater numbers were present or regular influxes of field mice from the mainland have allowed their pool of genetic possibilities to maintain the normal balance. What will be interesting is the impact the new mainland house mice now moving across the rubbish-filled causeway will have on this unique group of mammals, and whether, even now, we are watching the extinction of the legendary Bull Island Mouse.

No doubt the causeway will also lead to the arrival of such species as the fox and the stoat and perhaps recolonisation by the rabbit. This will provide interesting possibilities for the observant visitor. Hopefully we will not have to wait another 50 years for the next article on the Bull's mammals.

Chapter 17

THE BULL WALL

B. Healy *Department of Zoology University College Dublin*

The Bull Wall was constructed by piling boulders into a strip about twenty-five metres wide at the base. The 1.75 km section extending from the bridge to the monument is approximately seven metres high, the sea at high tide reaching about half way up the sloping sides. Beyond the monument, the wall continues seawards for almost another 1 km but at a lower level so that the whole wall is covered at high tide. The boulders used were mainly derived from Carboniferous limestone from the Clontarf area with some granite. The limestone merits close inspection because it is exceptionally rich in fossils with crinoids, (sea lilies), corals, and Bryozoa (moss animals) easily recognisable. (*Plate 17:1*) On the Island side of the wall, silting has occurred and the sand is at a higher level than on the Dublin side.

The loose structure of the sloping sides leaves spaces under and between the rocks into which the sea penetrates at high tide and which at low tide resemble miniature caves. All kinds of floating objects become trapped in these cavities including sand, shells, seaweed debris, sublittoral hydroids and Bryozoa and miscellaneous pieces of urban detritus. The moist, shady cavities with their decaying algal refuse, constitute an ideal habitat for many marine animals which find here a rich food supply and protection from the sun's rays and feeding birds. Unfortunately, these animals are often equally well protected from inquisitive naturalists and a serious analysis of the fauna is extremely difficult. During a recent preliminary survey, collections were made by lifting some of the smaller rocks and by reaching into the cavities and extracting as much material as possible by hand. Any estimation of population densities however is bound to be inaccurate and has not been attempted.

The fauna and flora are typical of rocky shores; no species of particular scientific interest has been found. Owing to the steep slope of the wall, the seaweed zones are contracted and the limits between them blurred. On the Dublin side of the main wall, a *Fucus spiralis* (flat wrack) zone marks the high water level with *Porphyra* (laver) seasonally abundant. Below this, *Fucus vesiculosus* (bladder wrack) is the dominant species while *Fucus serratus* (serrated wrack) forms a narrow band at the lower edge. On the island side, the upper and middle zones are similar but only isolated plants of *F. serratus* are found, sand having covered most of this zone. Green algae (*Ulva, Enteromorpha* and *Cladophora*) are common at all levels, especially on the upper surfaces of rocks and they often extend above the *F. spiralis* zone. The splash zone is represented by lichens, chiefly the orange *Caloplaca* with some *Verrucaria* (black) and *Xanthoria* (yellow). Some flowering plants, especially *Cochlearia,* grow in crevices at the top of the wall but there is rarely enough roothold for these to extend far.

Barnacles (*Balanus balanoides*) are found at all levels where the rocks are free from algae and beadlet anemones (*Actinia equina*) are frequent in shady places. Dog whelks (*Nucella lapillus*) which feed on barnacles and limpets (*Patella vulgata*) which are algal grazers are seen mainly in the lower half of the algal zone. The limpets are exceptionally large and tall, possibly due to their sheltered position among rocks. At the lower level are found large numbers of mussels (*Mytilus edulis*), particularly on the Dublin side. There is a poor representation of winkles—perhaps they fall into crevices. At the base of the wall where rocks are partly covered by sand, cockles (*Cardium edule*) and Polychaetes (*Cinatulus cinatus* and *Capitella capitata*) are often numerous. The latter may be present also among rocks where pockets of sand and seaweed debris collect. Rocks lying in the splash zone harbour a terrestrial fauna consisting of insects and woodlice, including the large slater *Ligia oceanica* which is characteristic of rocky habitats above high water. Rats are plentiful and unusually bold and must be important predators of both terrestrial and marine animals.

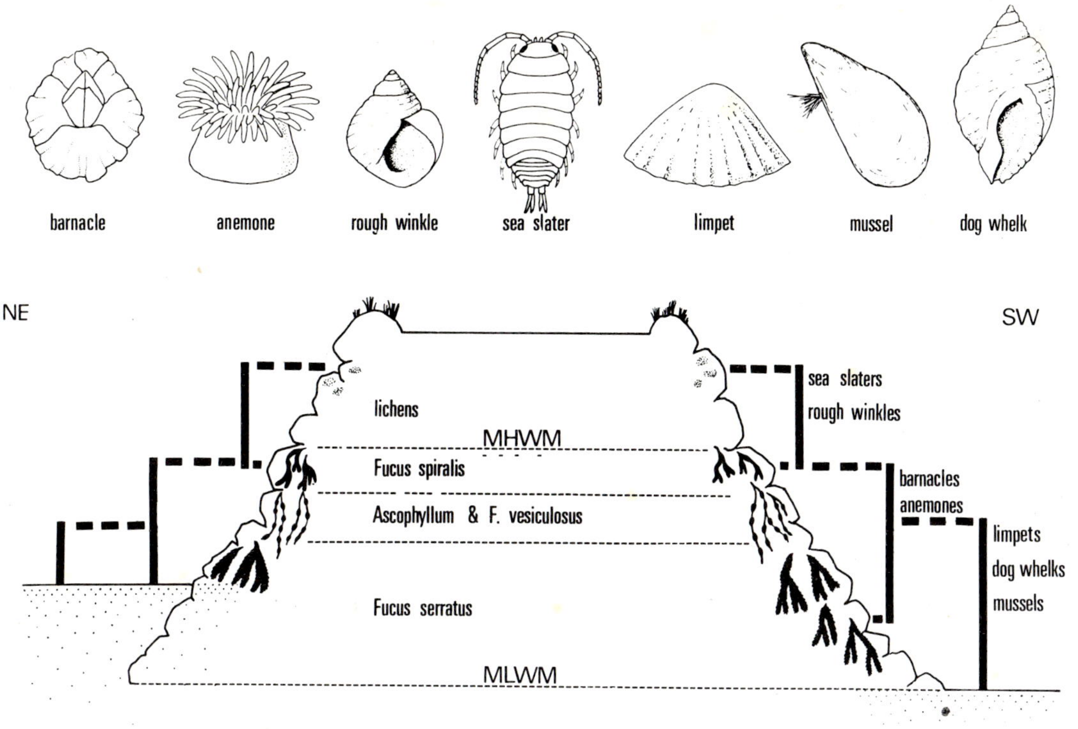

Fig. 17:1. Diagramatic section through the Bull Wall to show zonation of the dominant seaweeds and animals. Silting on the N.E. side has covered much of the level which corresponds to the lower shore and animals typical of this zone are better represented on the S.W. side.

On the section of wall beyond the monument, flora and fauna are both more abundant and more varied, and the differences between the two sides are less marked. The dominant seaweed on the upper, horizontal surface is *Ascophyllum nodosum* (knotted wrack) with some *F. spiralis. F. vesiculosus* covers most of the sides with *Chondrus crispus* (carrageen moss) in shaded places. *F. serratus* forms a narrow but distinct zone on the Dublin side but is present only in isolated patches on the Island side. Green algae are less abundant than on the main wall. Mussels are the dominant animal on the lower half of both sides, the number of barnacles is consequently reduced. Hydroids (*Dynamena pumila, Obelia geniculata*) are frequent on weed and sponges (*Halichondria panicea*) may sometimes be found on the underside of rocks.

Even allowing for the lack of intensive study, the species lists (see Appendix) are short compared with those for other shores in Dublin Bay. This is mainly due to poor habitat diversity. The fauna consists almost entirely of attached species since there are no rock pools and sand or mud for burrowing species is scarce. The absence of a kelp (*Laminaria*) zone or even a well developed *F. serratus* zone is responsible for the poverty of red algae, sponges, Bryozoa and other types characteristic of the lower shore.

Plate 17:1. Washed together on the Carboniferous shallow sea floor of some 330 million years ago, these fossils are now seen to advantage in the blocks of limestone from which this part of the Bull Wall is built. They include cylindrical portions and more or less separated disc-like ossicles which once gave skeletal strength to the flexible stems of crinoids (sea lilies). Also conspicuous are fragments of lace bryozoans (fenestellids) which are a feature of rocks of this age in various parts of the world. Mr. M.C. Smyth, Chief Engineer of the Dublin Port and Docks Board kindly draws our attention to the origin of this material which was drawn in carts from limestone quarries in the Clontarf area.

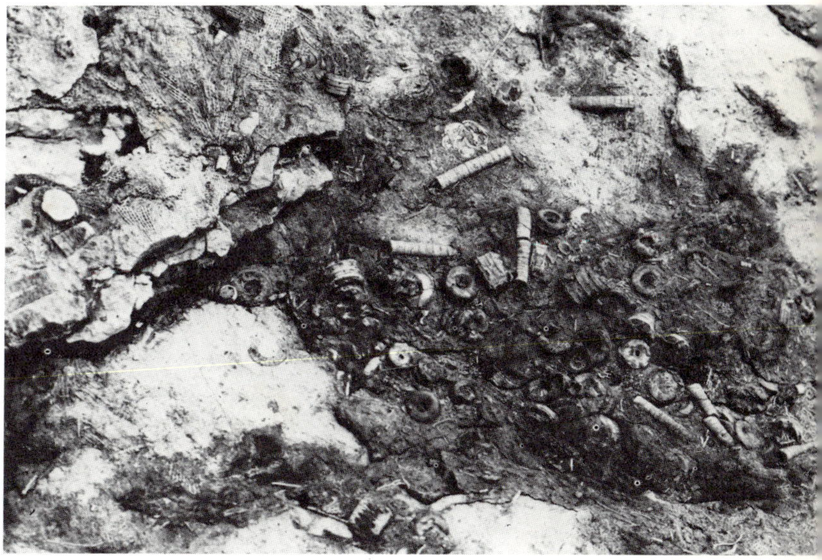

Acknowledgements

There has been no intensive study of either the flora or fauna of the Bull Wall. The preliminary survey on which the foregoing account is based was carried out by Declan Lyons and Bernhard Whelan of the Zoology Department, University College, Dublin, whose help is gratefully acknowledged.

Chapter 18

THE FUTURE OF THE ISLAND

J. S. Jackson *An Taisce—The National Trust for Ireland*

In this chapter the conservation of the North Bull Island is considered, raising the major issues and offering opinions on some of them. It is necessary to emphasise that in actual conservation, the hard facts relating to a particular case must be examined in detail before decisions are reached. Ideals and theoretical concepts may ultimately be combined in formulating a conservation policy, but the facts have a habit of reasserting themselves if ignored.

The importance of the North Bull Island lies in the unique combination of its values and in its immediate proximity to a densely populated area. Its pre-eminent recreational or amenity value to the citizens of Dublin, now in excess of 800,000, cannot be overemphasised. It is also a bird sanctuary of international importance, particularly for waders and wild fowl, covering some 1,404 hectares (3,470 st. acres), and including the lagoon, mudflats, saltmarsh, dunes, foreshore and offshore areas. All these habitats are of great ecological importance as an outdoor laboratory for the use of all, but particularly the post-primary and third level students of Dublin city and county.

Legislation

Environmental and nature conservation, in the absence of adequate legislation and appropriate land ownership, is reduced to a pious hope – an exercise in futility. The existing legislation which is relevant to the conservation of the North Bull Island includes:

(i) The Protection of Birds Act, 1930 under which initially the Minister for Justice, and later the Minister for Lands, have annually made ministerial orders declaring the North Bull Island, its lagoon and mudflats to be a bird sanctuary.

(ii) The Local Government (Planning and Development) Act, 1963, and particularly Sections 42 and 43 under which Dublin Corporation has made a Special Amenity Area Order in the context of Dublin Bay and including the North Bull Island. It should be noted that such an Order recognises the 'natural beauty', the scenery, amenity and recreational importance of the island; it confers no protection on the scientific values which are outside the scope of these Sections. To ensure such protection within the provisions of this Act a Conservation Order would have to be made in accordance with Section 46.

(iii) Section 15 of the Wildlife Bill, 1974 is concerned with Reserves and Refuges, the establishment of nature reserves by 'establishment order' on State lands and by 'recognition order' on lands not in State ownership. Although this Bill has obvious relevance to the North Bull Island it has not yet been enacted and its ultimate form and its effectiveness as an instrument of conservation are as yet unknown.

(iv) The Foreshore Act, 1933 is of little relevance as the Dublin Corporation's title of its island property extends down to the low water mark and the land between this and high water mark is not therefore vested in the Minister for Transport and Power.

(v) The bye-laws made by Dublin Corporation and operating on the North Bull Island are in accordance with the Open Spaces Act, 1906.

The degree of protection extended by legislation to the North Bull Island is essentially discretionary in the sense that an annual Ministerial Order is required to protect the bird sanctuary and maintain its status and a Special Amenity Area Order, made by Dublin Corporation in respect of Dublin Bay, a sub-section of which relates to the North Bull Island, requires to be renewed every five years in the quinquennial revision of the City plan. If the Ministerial Order is not renewed the bird sanctuary ceases to exist; if the revised City plan is varied so that the special amenity area status is not renewed, then that degree of environmental protection is withdrawn.

In order to ensure for posterity the continued existence of the bird sanctuary and to effectively protect the island as an amenity area and, in part, as a nature reserve, it is essential that legislation,

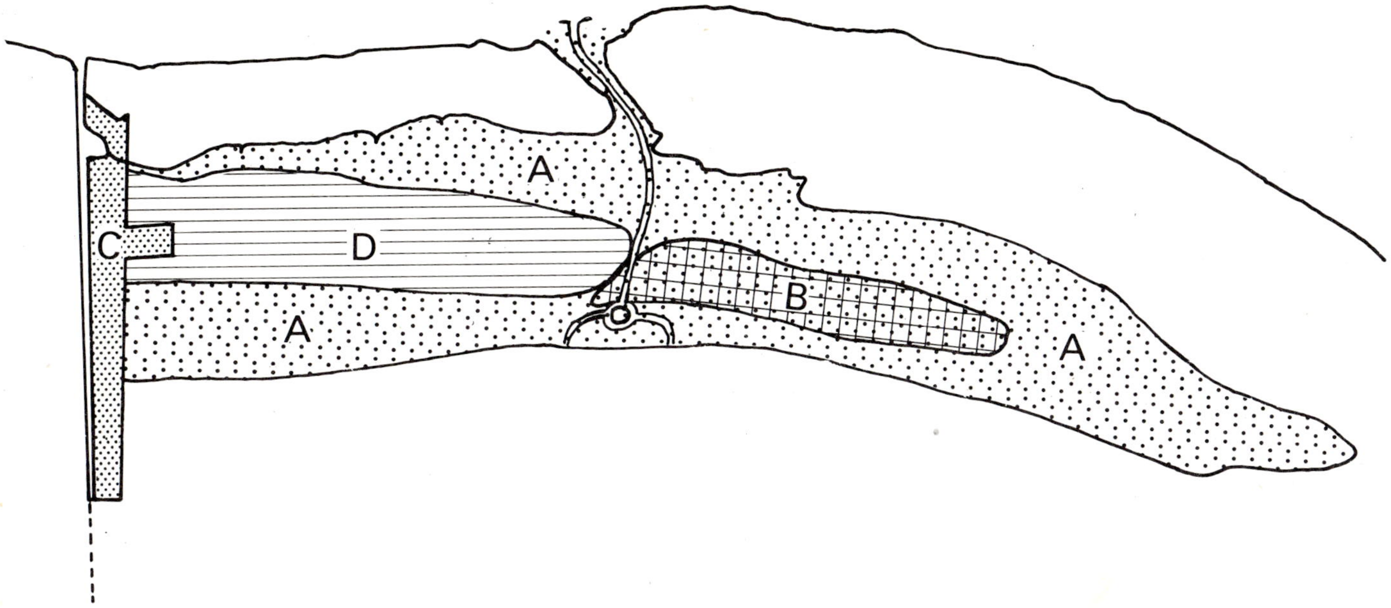

Fig. 18:1. Land ownership map. A. Dublin Corporation. B. Leased by St. Anne's Golf Club from Dublin Corporation. C. Dublin Port and Docks Board. D. Royal Dublin Golf Club.

specifically drafted for the North Bull Island, be enacted. Comparable legislation already exists in respect of three other specific areas which are of importance as recreational amenities, the relevant legislation for two being enacted since the formation of the State, viz. The Saint Stephen's Green (Dublin) Act, 1877, the Phoenix Park Act, 1925 and the Bourn Vincent Memorial Park Act, 1932. Similar specific legislation is now required for the North Bull Island which would probably involve vesting in the State that part of the island currently owned by Dublin Corporation. This would not conflict in any way with the aims of the Corporation for the future of the island nor should this recommendation be interpreted as in any way critical of the role of Dublin Corporation as a planning agency with conservation responsibilities.

Ownership

Ownership of the island originated in royal charter granted by Charles II to the Vernon Estate and related to tidal flats and other sections of the sea floor on the north side of Dublin Bay. In the same way such 'land' on the south side of the Bay was granted to the Pembroke Estate. Currently three owners are involved in the ownership of the North Bull Island (*See Fig. 18:1*).

(i) The Royal Dublin Golf Club which owns its links covering some 165 st. acres (66.7 ha);

(ii) Dublin Port and Docks Board which owns a strip of land contiguous to the North Bull Wall which was acquired prior to the construction of the sea-wall in 1820 and which covers about 47 st. acres (19 ha);

(iii) Dublin Corporation who owns the remainder of the island down to low water mark. The area increases annually by island accretion. In 1960 the total area of the island to high water mark was estimated to be 701 st. acres (283.6 ha) of which Dublin Corporation owned some 489 acres (197.8 ha). St. Anne's Golf Club lease their 64½ acre (26.1 ha) links from the Corporation.

Educational and Scientific values

Post-primary biology students

The importance of Bull Island as a study area has increased enormously since the introduction of biological projects into post-primary schools and since biology was introduced as a subject in the

Leaving Certificate curriculum in 1971. Prior to that date biological sciences in the schools included botany, physiology and hygiene but these were essentially non-field and non-laboratory subjects and lacking an ecological basis. This course attracted very few boys.

It can be appreciated from *Table 18:1* that within a period of five years biology has become a major subject in post-primary schools, and taken by about one third of the candidates in the country.

	1971	1975
Leaving Biology	3,757 (15%)	10,765 (32%)
Intermediate Biology	2,063	3,101
Leaving Agricultural Science	750	796
Total:	6,570	14,662

Table 18:1 Post-primary students studying courses with field biology content 1971–1975.

In the school year 1975–76 there were 72,480 post-primary students located in the 182 schools of Dublin City and County, all being within a convenient or reasonable distance of Bull Island. The author predicts that some 9,900 will sit for Leaving Certificate examinations in 1977. Of this number a third (approximately 3,300 students) will take biology as a subject. With students in geography additional to these numbers, the island is clearly a site of great actual and potential importance as a demonstration and study area for school children.

Third level education

In any one year there are over 1,000 University students in Dublin whose degree courses in biology and earth sciences can be associated with the island. These range from excursions in which perhaps 100 students spend half a day being shown a series of habitats, to the three years spent by research students training in a particular area of science. The text of this book is largely based on the work of students and staff of the Universities, indicating again that the combination of quality of environment and proximity to the city is important.

Education of the general public

The island may readily play a role in formal education because of the availability of informed guidance from school teachers and university staff. This is not the situation faced by the lay person who wishes to learn about the island. Supply of information to the public is vital for the adequate future conservation of all the values discussed. At its most basic, the public will be footing the bill for all conservation costs entailed. There is also a self-sustaining demand for information on natural history topics.

Apart from publications such as this, two major information channels could be used effectively. *An interpretative centre* could be the physical base for dissemination of information and literature, staffed either by suitably trained permanent personnel or a panel of volunteer naturalists. Literature designed perhaps on a seasonal basis or with regard to particular zones would make the public aware of both natural history and the need to manage the site in particular ways.

Nature trails are now familiar in inland situations and are welcomed by visitors for the interest added to a simple walk. It is probable that one or more trails could be designed for the island, but it is clear because of the fragile nature of dunes and saltmarsh and the high visitor numbers anticipated, that great care would be required.

Scientific Research

In most of the preceding chapters, reference has been made to scientific work carried out in the past, currently in progress or gaps to be filled in the future. These chapters document the importance of the Bull as a major national scientific resource, actively in use.

There are international responsibilities too of having in this city bird sanctuary the greatest density in Europe of overwintering waders and wild fowl.

Photo: B. Kinsel

The recreational role

The vital importance of coastal sites in the Irish recreational context has recently been established formally and re-emphasised in a number of studies. The amenity or recreational importance of the Bull Island particularly at weekends, to the residents of Dublin city north of the Liffey cannot be overstressed. In a recent survey of outdoor recreational activities Mawhinney (1975) has established, through interviews with 1,336 people in the Dublin City and County area, that 'day trips to the coast' were the most popular outdoor recreational activity which those interviewed 'said they did at weekends in Summer' and represented 35 per cent of the sample. This particular type of recreation was second only to 'going for a walk' in activities which those interviewed reported doing both 'every weekend in Summer' and in 'activities participated in over the weekend previous to interview'. In the context of activities during weekday evenings a 'trip to the coast' was subordinate in popularity only to 'going for a walk' and 'swimming'.

In an analysis of age groups Mawhinney established that the highest levels of participation for trips to both country and coast were in the 30–39 year old age group, ('country' 42.2 per cent; 'coast' 45.2 per cent) and that this age group is associated with families of young children. The tabulated figures indicated the popularity of the day trip to the coast at weekends as a recreational activity undertaken by the family as a whole, in contrast to other activities which were participated in by only certain members of the family. Tabulated data also demonstrated that, up to a total of six, i.e. parents and four children, day trips to the coast increased in popularity with the size of the family, 22.9 per cent representing the 'family' of one, steadily increasing to 46.8 per cent for the family of six.

Of car owners interviewed 38 per cent said they went on day trips to the coast. It is significant, in the Dublin context, to note that in 'households with no car', 20.7 per cent went on day trips to the country whereas 27.1 per cent went on day trips to the coast. This is presumably due to the ready accessibility of beaches to Dubliners without cars and, sociologically, a fact which must not be overlooked in the context of Bull Island. This point is dramatically reinforced visually in Mawhinney's Map No. 2 – 'Visits to beaches, etc.' – in which the lines, each representing 'one person, one trip' are so concentrated and superimposed in the Bull Island area as to result in an almost completely black zone. It is also significant to note that the majority of the day trips to the island originate within a few miles of Dollymount. These points are emphasised in considering the mode of transport to the beach, Although the car accounted for 68.5 per cent, nevertheless appreciable numbers travelled by public transport, 15.1 per cent, and on foot, 13.7 per cent, 'a phenomenon which reflects the proximity of important beaches and swimming places to much of the population of the Greater Dublin area'. In this the Bull Island plays a significant part, a point reinforced by the fact that, in 'going to the beach', the travelling time for 19.2 per cent of those interviewed was less than ¼ hour and over 41 per cent travelled for less than half an hour.

There can be little doubt that the environmental conservation of the Bull Island as a scientific and recreational area will present daunting problems to Dublin Corporation planners in the immediate future. In projected population estimates Buchanan has predicted that, by 1986 and only a decade away, the population of the coastal areas will be 1,800,000 persons, a net increase over the data of the study of some 38 per cent and representing about 54 per cent of the total projected population of the Republic. Using these data in the 'Report on the Findings and Recommendations of the National Coastline Project: 1, A Strategy for Development and Conservation' the authors pointed out that 'the coastline whilst constituting only 5.5 per cent of the total land area of the Twenty-six Counties will be required to cater for more than half of the total population' and that 'clearly the coastal areas close to the major cities will come under the greatest pressure for development'.

Development strategy for the future

Assuming that the future role of the island implies the acceptance of a multi-use policy catering for recreational and scientific interests, it is essential that the conflict between these interests be reconciled, or minimised, by unambiguous planning. The two golf links are recreational areas to which the general public does not enjoy access. The boundary of the Royal Dublin golf links, whilst not aesthetically inspiring, nevertheless at least has the merit of avoiding ambiguity. The boundary of the area leased to St. Anne's golf club should be clearly defined on the ground by unobtrusive but unambiguous means.

Establishment of a nature reserve

The interpretation of 'nature reserve status' needs to be carefully considered. Basically it entails the regulation of access and activity within a defined area so that nature conservation, scientific research and education are facilitated.

This definition is clear and unambiguous in planning terms, but in terms of management the need for sensitive judgement is paramount. The degree of protection afforded to a nature reserve, and its size and location are all critical issues on the island.

The ecologist may argue cogently that the whole island is scientifically important and that the total area merits a nature reserve status. This view would obviously conflict with amenity or recreational interests which could present a case with equal cogency that an open space within the city boundaries must primarily be a amenity area in which the scientific issues are subordinate. The scientific importance of the island has already been discussed as also has the great number of post-primary students within the city and county area which could use the island for ecological demonstration and study. The solution to the dilemma is obviously one of compromise in which both interests are catered for and neither gets all it is seeking.

Although the total area of the island is of ecological interest the areas of the two golf links represent artificially modified ecosystems which, in any case, are not accessible to the general public or to large parties of students, and it is therefore the relatively undisturbed dune and spit area of the north of the island together with the salt marsh and lagoon which are of prime scientific importance. It is suggested therefore that a nature reserve status be conferred on the lagoonal area, the saltmarsh and the dune area north of St. Anne's golf links.

The great bulk of visitors using the island tend to congregate along the east margin of the island, along the foreshore and in the contiguous dune area, and virtually all of this, with the exception of the north of the island, would lie outside this recommended nature reserve. It is recommended that the public have access to part of the saltmarsh in order that its unique character may be appreciated. An area adjoining the access road from the causeway would be most suitable from the point of view of easy access and involving least disturbance to bird feeding and roosting areas and to long-term experimental study areas.

It is essential, if long-term scientific field studies in ecology and biology are to be conducted in any meaningful way, that adequate protection must be afforded to the nature reserve.

There are many devices by which protection is made possible. The most essential are a combination of the unambiguous marking of boundaries and the explanation to the public of the significance of nature reserve status. Distance from car parks is as useful a device as fencing in regulating the public pressures on an area. The development of a long-term management plan for any Bull Island Nature Reserve will mean evaluating many items of detail. Hopefully, a flexible and experimental approach will be employed.

The dune area

The dune area should be open to public access as far north as the northern boundary of the St. Anne's links but, north of that point, the dune area, in being relatively undisturbed and unmodified and in possessing a dune-slack (the Alder Marsh) of unusual ecological interest, is of great ecological importance and should be designated part of the proposed nature reserve.

Intensive public use of dunes presents difficult practical management problems. The beach and dunes are geomorphologically interrelated. Compaction of beach sand may reduce dune growth, destruction or erosion of dunes may both alter the beach profile or lead to large quantities of mobile sand to be present just above high tide mark. Golf courses or service roads would also be affected by deep blown sand. The principal enemies of dune ridges are destruction of *Ammophila* by intensive and prolonged public pressure, and burning. Where this has already happened, close to the Bull Wall, management to encourage a dune pasture based on *Festuca rubra* might be considered. Such turf is an eminently suitable surface for picnics, and can support greater human activity. The ecologist can have an important part to play in improving the recreational amenities of such areas.

The beach

The beach should be open to public access but cars should be totally excluded from this area. There is palpable conflict in the recreational use of the beach and its use by motor cars in that young children and elderly people together with those who walk to the beach or travel by public transport tend to congregate close to parking facilities or to the public transport termini; they therefore tend to use the beach in the vicinity of the North Bull Wall and of the new main car park. These, however, are the access points used by motorists to drive onto the beach, seriously militating against relaxation of parents of young children, of the timid and of the elderly. This is a problem which will become exacerbated in future years and, in this context, the planner must be aware of the implied warning in the National Coastline Study (p. 30) " . . . it is clear that daytripping is a major form of recreational activity . . . and a major 'demand' factor to be reckoned with . . . A conservative estimate of domestic tripping to coastal areas in 1970 suggests that between four and a half million and five million 'person' trips (1.5 million car trips approximately) were made in that year. Of these approximately 47% occurred in the three catchment areas serving the east coast from Carlingford to Arklow (a coastal zone including Bull Island). By the year 1986, (now only a decade away), projected figures estimate that this will have increased to 7,520,000 person trips". Therefore the time to make a policy decision of excluding cars from the beach on Bull Island is now, not in ten years' time when the problem will be extremely serious.

The beach is increasingly used by barbecue parties; barbecue areas, with shallow ash pits and robustly constructed grills, should be provided, thereby protecting the beach against the indiscriminate siting of fires and the contamination of beach sand by charcoal and ash.

It is not generally recognised by planners that beaches have an ecological limitation in recreational use as well as a dimensional, aesthetic or human tolerance aspect. The beach is regarded by many planners as an enormous filter bed in which aerobic bacterial activity decomposes organic material introduced into it; it is self-cleansing. However, even sand filters can be overloaded and if, by over-use by too many people, the amount of discarded litter incorporated into the sand seriously reduces pore space and, as a consequence, the ability of the sand to self-cleanse, and concurrently appreciable quantities of organic material requiring decomposition are introduced into the sand, then aerobic breakdown can reach its limit and anaerobic conditions can be initiated, resulting in a black sulphurous zone developing between water table and beach surface and degradation of the beach environment results. When children dig into the beach this aesthetically objectionable and malodorous zone is quickly intersected.

It is however frequently asserted in planning documents that the capacity of a beach to support recreational activity is limited only by such factors as the density tolerance of users, or the capacity of car parks.

Although these observations are valid in terms of planning for beaches subject to moderate recreational use they are unacceptable, in the context of gross over-use, to ecologically orientated environmental conservationists. The potential recreational capacity of a beach is finite and, because of its proximity to the population of Dublin and because of its popularity, the Bull Island will be inevitably exposed to tremendous recreational pressure. It is therefore of value to examine published research work on the recreational capacity of beaches, a subject in which very little study has been undertaken, and for which available data are meagre.

Densities which beaches can tolerate in a planning sense vary from the pragmatic figure for Brittas Bay of 250 persons per quarter hectare, or one person per 10 square metres (approximately 3.3 m square), (Mawhinney 1971b), to as much as 4,000 persons per hectare, or one person per 2.5 square metres (approximately 1.6 m square), (data from the Netherlands quoted in the above study).

If the Bull Island were to accommodate visitors on, say, 3 km of the beach, at a density of 4,000 persons per hectare and assuming an average beach width of 60 m, this would amount to 72,000 people. Research has shown that 68.5 per cent of Dubliners travel to the coast by car and if this figure is accepted for the Bull Island it would represent 49,000 people out of the total of 72,000. At an average passenger load of four per car some 12,330 cars would require parking facilities which, at a density of 371 cars per hectare, would require some 33.23 hectares or 82.1 statute acres. At the lower density figure of 250 persons per quarter hectare of beach the above figures would be quartered and some 20.5 statute acres would be required for parking.

As the National Coastline Project has noted, the density use of a beach is controlled by the available parking space and a policy decision will have to be taken by Dublin Corporation regarding this density. Whether the higher or lower density figure is adopted the parking space required is unacceptably high for the limited area of the island open to the public and, although an appreciable number of cars may be accommodated in strategically sited car parks on the island, it is recommended that parking facilities be made available on the mainland and a public transport feeder service provided during the Summer weekend periods to ferry visitors from their cars to and from the beach.

Car parking areas

Dublin Corporation has tentatively suggested siting a number of car parks at the eastern margin of the dune area between the North Bull Wall and the new main car park serviced by the causeway. This is to be recommended and in my opinion this zone, almost a mile in length, should be developed intensively for car parking and picnicing facilities. The main car park serviced by the causeway could be enlarged. Any additional car parking should be accommodated on the mainland with, at summer weekends, a public transport feeder service to and from the beach. No car parks should be sited to the north of the main car park and the more energetic visitors should be encouraged to use this zone for recreation.

Roads

A road linking the North Bull Wall to the main car park should be created in order to service the proposed new car parking facilities. The Corporation should consider a one-way traffic system for private cars, with access to the island *via* the North Bull Wall and egress *via* the causeway. Buses cannot use the wooden bridge connecting the mainland at Dollymount to the North Bull Wall and these should have a separate lane on the causeway route to facilitate movement between the mainland and the island and particularly on summer weekends when the island receives maximum visitors and when the proposed feeder service would be in operation.

Acknowledgements

Many people gave unstinted help in the preparation of this chapter through discussion and in answering questions. I would specifically like to thank Mr. Davy Byrne, Dublin Corporation (Community and Environment Department) who made available to me several maps showing the ownership pattern on the island. I would also like to thank Mr. Kevin McDonagh and Mr. U. Wrynn of the Department of Education who discussed the post-primary situation in Dublin City and County with me and who supplied statistical data used in the above text. And finally I would like to thank the General Editor whose tact and tenacity have made my contribution possible.

FURTHER READING

This list has been compiled by authors to guide the reader in the literature of their subject. This includes the necessarily complex taxonomic literature associated with the check lists. If a general work refers directly to the Bull Island, it has been included in the bibliography.

An Foras Forbartha (1968). *Regional Studies in Ireland.* An Foras Forbartha, Dublin.

An Foras Forbartha and Bord Fáilte Éireann (1972–73). National Coastline Study. *Vol.* **1** A national strategy for development and conservation (1972). *Vol.* **2** County Reports: Notes on development and strategy (1973). *Vol.* **3** A procedural manual (1973) An Foras Forbartha, Dublin.

Angel, H. (1974). *The World of an Estuary.* Faber & Faber, London.

Barnes, R. S. K. & Green, J. (eds.) (1972). *The Estuarine Environment.* Applied Science Publishers Ltd., London.

Barnes, R. S. K. (1974). *Estuarine Biology,* Arnold, London.

Barrett, J. H. & Yonge, C. M. (1958). *Collins Pocket Guide to the Seashore.* Collins, London.

Bird, E. C. F. (1969). *An Introduction to Systematic Geomorphology v.* **4**. *Coasts.* The M.I.T. Press, Cambridge, Mass. & London.

Bliding, C. (1963). A critical survey of European taxa in Ulvales. Part I. *Capsosiphon, Percursaria, Blidingia, Enteromorpha. Op. bot. Soc. bot. Lund.* **8**, 1–160.

Buchsbaum, R (1951). *Animals Without Backbones* 2 vols. Penguin Books, London.

Burrows, E. M. & Lodge, S. (1951). Autecology and the species problem in *Fucus. J. mar. biol. Ass. U.K.* **30**, 161–176.

Chapman, V. J. (1939). Studies in salt-marsh ecology sections IV and V. *J. Ecol.* **27**, 160–201.

Chapman, V. J. (1960). *Salt-marshes and Salt Deserts of the World.* Leonard Hill, London.

Cummins, H. (1930). Experiments on establishments of Rice Grass in the Lee. *Econ. Proc. R. Dubl. Soc.* **2**, 419–421.

Dennis, R. W. G. (1960). *British Cup Fungi and their Allies.* The Royal Society.

Eltringham, S. K. (1971). *Life in Mud and Sand.* English Universities Press, London.

Goodman, P. J., Braybrooks, E. M., Lambert, J. M. & Marchant, C. J. (1969). Spartina Schreb. *J. Ecol.* **57**, 285–313.

Green J. (1968). *The Biology of Estuarine Animals.* Sidgwick and Jackson, London.

Hepburn, I. (1952). *Flowers of the Coast.* Collins, London.

Hudson, R. (1971). *A Species List of British and Irish Birds.* Guide 13. British Trust for Ornithology, Tring.

Hubbard, C. E. (1957). In Report of B. E. S. Symposium on *Spartina. J. Ecol.* **45**, 33–35.

Hubbard, C. E. (1968). *Grasses,* 2nd edn. Penguin Books, London.

James, P.W . (1965). A new check list of British lichens. *Lichenologist,* **3**, 95–153.

King, C. A. M. (1972). *Beaches and Coasts.* Edward Arnold, London.

Mawhinney, K. A. (1971a). *Development in the Brittas Bay Area, Co. Wicklow.* An Foras Forbartha.

Mawhinney, K. A. (1971b). Brittas Bay Study – Part 2. Conservation and recreational use of the beach and dunes. Draft report. An Foras Forbartha.

Mobberley, D. G. (1956) Taxonomy and distribution of the genus *Spartina. Iowa St. J.Sci.* **30**, 471–574.

Nelson-Smith, A. (1972). *Oil Pollution and Marine Ecology.* Elek Science, London.

Nicol, E. A. T. (1935). The ecology of a salt-marsh. *J. mar. biol. Ass. U.K.* **20** N.S.2, 203–261.

Patten, C. J. (1906). *The Aquatic Birds of Great Britain and Ireland.* London, R. H. Porter.

Paton, J. A. (1965). *Census Catalogue of British Hepatics.* ed. 4. Br. bryol. Soc.

Parke, M. & Dixon, P. S. (1968). Check list of British marine algae—second revision. *J. mar. biol. Ass. U.K.* **48**, 783–832.

Pettijohn, F. J., Potter, P. E. & Siever, R. (1972). *Sand and Sandstone.* Springer-Verlag, Berlin, Heidelberg and New York.

Quinn, A. C. M. (1971). *Sand-dunes, Formation, Erosion and Management.* An Foras Forbartha.

Ranwell, D. S. (1967). World Resources of *Spartina townsendii* sensu late and Economic uses of *Spartina* Marshland. *J. Appl. Ecol.* **4**, 239–256.

Ranwell, D. S. (1972). *Ecology of Salt-marshes and Sand-dunes.* Chapman and Hall, London.

Rea, C. (1922). *British Basidiomycetae.* Cambridge University Press, London.

Reineck, H. E. & Singh, I. B. (1973). *Depositional Sedimentary Environments* Springer-Verlag, Berlin, Heidelberg, & New York.

Salisbury, E. J. (1952). *Downs and Dunes.* G. Bell & Sons, London.

Schäfer, W. (1972). *Ecology and Palaeoecology of marine environments.* Oliver & Boyd, Edinburgh.

Singer, R. (1962). *The Agaricales in modern taxonomy.* 2nd edn. J. Cramer, Weinheim.

Stapf, O. (1908). *Spartina townsendii.* Gdnr's Chron. 43, 33–35.

Steers, J. A. (ed.) (1934). *Scolt Head Island,* W. Heffer, & Sons, Cambridge.

Steers, J. A. (1953). *The Sea Coast.* Collins, London.

Tebbel, N. (1966). *British Bivalve Seashells.* British Museum (Natural History).

Thorson, G. (1971). *Life in the Sea.* Weidenfeld and Nicholson, London.

Yonge, C. M. (1966). *The Sea Shore.* Collins, London.

Verwey, J., Morzer Bruyns, M. F. & Gorter, H. P. (1966). *A plea for the Wadden Sea.* Contact Committee for Nature and Landscape Preservation. Herengracht 540, Amsterdam, Netherlands.

Warburg, E. F. (1963). *Census Catalogue of British Mosses,* ed. 3. Br. Bryol. Soc.

ADDITIONAL FURTHER READING

Campbell, A. C. and Nicholls, J. (1976)
The Hamlyn Guide to the Sea Shore and shallow seas of Britain and Europe. Hamlyn, London.

Colgan, N. (1914)
The Opisthobranch fauna of the shores and shallow waters of Dublin Bay. *Ir. Nat.,* 23, 161–204.
1930 The marine Mollusca of the shores and shallow waters of Co. Dublin. *Proc. R. Ir. Acad.,* *39*. 391–424.

Fairley, J. S. (1975)
An Irish beast book, Blackstaff Press, Belfast.

Kloet, G. S. and Hincks, W. D. (1945)
A check list of British insects. Kloet & Hincks. Stockport.
(1964) A check list of British insects 2nd edition revised. Part 1. Small orders and Hemiptera. Royal Entomological Society, London.
(1972) Part 2. Lepidoptera. Royal Entomological Society, London.

Nichols, A. R. (1900)
A list of the marine Mollusca of Ireland. *Proc. R. Ir. Acad.,* 3rd Ser., *5*, 477–662.

O'Riordan, C. (1965)
A catalogue of the collection of Irish fishes in the National Museum of Ireland. Dublin.

O'Rourke, F. J. (1970)
The fauna of Ireland. Mercier Press, Cork.

Southern, R. (1910)
The marine worms (Annelida) of Dublin Bay and the adjoining district. *Proc. R. Ir. Acad., 28,* 215–246.
1936 Turbellaria of Ireland. *Proc. R. Ir. Acad., 43*B, 43–72.

Webb, D. A. (1967)
An Irish Flora, 5th Edition. Dundalgan Press, Dundalk.

BIBLIOGRAPHY

Abdel-Moetz, M. K. (1957). *The marine annelid fauna of Dublin Bay and neighbouring waters.* Ph.D. thesis, University College, Dublin.

Anon (1974). Report on wetlands of international and national importance in the Republic of Ireland. Forest and Wildlife Service, Department of Lands, Dublin. (Unpublished).

Barrington, R. M. (1908). List of Dublin and Wicklow birds. In Cole and Praeger (1908), 113–129.

Blake-Knox, H. (1866). The migratory and wandering birds of the County Dublin with the times of their arrival and departure. *Zoologist* **24,** 220–227, 295–307, 479–483.

Boyle, P. J. (1972). *Spartina in Ireland.* Unpublished PhD thesis, University College, Dublin.

Boyle, P. J. (1976). A variant *Spartina* in three regions north of Dublin. *Sci. Proc. R. Dubl. Soc.* A.**5**.

Brandt, E. (1971). *The ecological distribution of the littoral Isopods in Dublin Bay.* Moderatorship thesis, Department of Zoology, University of Dublin, Trinity College.

Broughton, R. et al. (1801). *Accounts, etc., Presented to the House of Commons relating to Inland Navigations of Ireland and the Port of Dublin.* London, July, 1806.

Cabot, D. (1966). *A Guide on How and Where to Watch Wildlife.* Royal Dublin Society.

Caffrey, J. M. (1975). *The ecological life history of sand dune winter annuals.* Unpublished project report, Department of Botany, University College, Dublin.

Cole, G. A. J. & Praeger, R. L. (Eds.) (1908). *Handbook to the Dublin District.* British Association, Dublin.

Denneny, J. M. (1951). *The vegetation of the salt marsh, N. Bull Island. Co. Dublin.* Unpublished project report, Department of Botany, University College, Dublin.

Doyle, J. (1934). *Spartina townsendii* H. & J. Groves at North Bull, Co. Dublin. *Ir. Nat. J.* **5**, 158.

Fairley, J. (1971). A critical reappraisal of the status in Ireland of the eastern house mouse, *Mus musculus orientalis*. Cretzmar. *Ir. Nat. J.* **17**, 2–5.

Farrington, A. & Haughton, J. P. (1947). The North Bull Island, Co. Dublin. *Ir. Nat. J.* **9**, 46–48.

Fenner, M. (1971). *The comparative ecology of two salt marsh grasses.* Unpublished Moderatorship thesis, Department of Botany, University of Dublin, Trinity College.

Finlay, W, D., Jackson, J. S., Webb, D. A. & West, A. E. J. (1963). A preliminary report from the subcommittee on nature conservation. An Taisce report. Unpublished.

Flood, D. T. (1975). The Birth of the Bull Island. *Dubl. Hist. Rec.* **28**, 142–153.

Griffin, P. M. (1969). *A vegetation survey of the North-East half of North Bull Island, Co. Dublin.* Unpublished project report, Department of Botany, University College, Dublin.

Griffith, J. P. (1879). The Improvement of the Bar of Dublin Harbour by Artificial Scour.*Proc. Inst. Civ. Engrs.,* **58**, 104–145.

Halbert, J. N. (1920). The Acarina of the sea shore. *Proc. R. Ir. Acad.* **35B**, 106–152

Halbert, J. N. (1935). A list of Irish Hemiptera (Heteroptera and Cicadina). *Proc. R. Ir. Acad.* **42B**, 211–311.

Harris, C. R. (1973). *The Evolution of North Bull Island, Dublin Bay.* M.Sc. thesis, University of Dublin, Trinity College.

Harris, C. R. (1974). The Evolution of North Bull Island, Dublin Bay. *Sci. Proc. R. Dubl. Soc. A,* **5**, 237–252.

Harris, C. R. (1974). A quantitative study of compaction with depth in a vesicular sand layer in an intertidal area of Dublin Bay, Ireland. *Jour. Sediment. Petrol.* **44**, 1024–1028.

Hassel, F. C. (1953). The Fungi of North Bull Island, Co. Dublin. *Ir. Nat. J.* **11**, 1–5.

Hassel, F. C. (1959). Additions to the fungi from the North Bull Island, Co. Dublin. *Ir. Nat. J.* **13**, 86.

Healy, B. (1975). Fauna of the salt-marsh, North Bull Island, Dublin. *Proc. R. Ir. Acad.* **75B**, 225-244.

Hutchinson, C. D. & Keys, J. M. (1973). The numbers of wildfowl on the North Bull Island, Co. Dublin. *Irish Bird Report,* **20**, 35–43.

Hutchinson, C. D. & Rochford, J. M. (1974). The numbers of waders on the North Bull Island, Co. Dublin. *Irish Bird Report,* **21**, 68–77.

Hutchinson, C. D. (ed.) (1975). *The Birds of Dublin and Wicklow.* Irish Wildbird Conservancy, Dublin.

Jackson, J. S. (1965). Excursion guide to the Bull Island. Museums Association Conference. Unpublished.

Jackson, J. S., Went, A. E. J. & Watts, W. A. (1965). Memorandum from An Taisce, the National Trust for Ireland, to the planning and development committee of Dublin Corporation. Unpublished.

Jameson, H. L. (1898). On the probable case of protective coloration in the house mouse (Mus musculus Linn.) *J. Lin. Soc. Zool.,* **26**, 456–473.

Jeffrey, D. W., Pitkin, P. H., & West, A. B. (in Press). Intertidal environment of northern Dublin Bay.

Kennedy, P. G. (1935–1936). Bird life on the North Bull. *Ir. Nat. J.,* **5**, 165–168, 190–192, 219–223, 260–264, 271–274, 299–302; **6**, 12–16, 46–48, 68–72, 88–92, 120–125.

Kennedy, P. G. (1947). Additional notes on the bird life of the North Bull. *Ir. Nat. J.,* **9**, 1–9.

Kennedy, P. G. (1949). *Violation of Sanctuary.* Studies **38**, 37–45.

Kennedy, P. G. (1953). *An Irish Santuary – Birds of the North Bull.* Three Candles Press, Dublin,

Kennedy, P. G., Ruttledge, R. F. & Scroope, C. F. (1954). *Birds of Ireland.* Oliver & Boyd, Edinburgh & London.

Kunz, Y. (1969). Salinity preference and salinity tolerance of the katadromous teleost *Pomatoschistus (Gobius) microps* Kroyer. *Proc. R. Ir. Acad.,* **68B**, 77–88.

Lambe, E. (1967). *A comparative assessment of some phytosociological methods.* Unpublished project report, Department of Botany, University College, Dublin.

Letts, E. A. & Adeney, W. E. (1907). *Pollution of Estuaries and Tidal Waters.* Report of Royal Commission on Sewage Disposal, 141–168.

Max, M. D., Harris, C. R., Geoghehan, M. A., Cathcart, G. S., Ní Chonchuir, M. E., & Fahy, G. J. (in press). Preliminary report on the marine geology of the approaches to Dublin. Report Series 75/5, Geol. Surv. Ireland.

McGrath, D. (1972). *Studies on the Oligochaetes of Dublin Bay.* Unpublished project report, Department of Zoology, University College, Dublin.

McGrath, D. (1975). Notes on some Irish Marine Littoral and fresh water Oligochaeta (Annelida). *Ir. Nat. J.* **18**, 216–218.

McNamee, K. (1976). *An Experimental Study of an Irish Salt-marsh.* Unpublished Ph.D. thesis, Department of Botany, University of Dublin, Trinity College.

Malone, R. (1970). *A study of direct gradient analysis as applied to salt-marsh vegetation.* Unpublished project report, Department of Botany, University College, Dublin.

Mawhinney, K. A. (1975). *Survey of recreational activities in Dublin City and County.* An Foras Forbartha. Unpublished typescript.

Mitchell, G. F. (1972). The Pleistocene history of the Irish Sea: Second Approximation. *Sci. Proc. R. Dubl. Soc. A,* **4**, 181–199.

Moriarty, C. (1955). Behaviour of hares on mud flats. *Ir. Nat. J.* **11**, 310.

Moriarty, C. (1971). The North Bull. *Irish Times,* 10–11 February, 1976.

Moriarty, C. (1976). Bull Island Birds. *Fieldfare*, 75, Dublin Naturalists' Field Club.

Ní Annrachain, O. (1972). *A phytosociological survey and mapping of the southern half of the North Bull Island.* Unpublished project report, Department of Botany, University College, Dublin.

O'Connor, P. (1936). A Contribution to knowledge of the Irish fungi. *Sci. Proc. R. Dubl. Soc.* **21**.

O'Connor, P. (1949). A further contribution to knowledge of Irish fungi. *Sci. Proc. R. Dubl.*, **25**, 33–53.

O'Gorman, F. & Wymes, E. (1973). *The future of Irish wildlife – a blueprint for development.* An Foras Taluntais, Dublin.

O'Gorman, F. & von Rezzori, A. (1976). The Bull Island Mouse *Mus musculus jamesonii* Krausse. A reappraisal of its status. (in press).

O'Mahoney, E. G. (1924). Records for *Cholera nigrita, Cassida nobilis* and other beetles in Co. Dublin. *Ir. Nat.* **33**, 4.

O'Mahoney, E. G. (1924). *Athous difformis* Lac. and other beetles in Co. Dublin *Ir. Nat.* **33**, 121–122.

O'Mahony, E. (1931). Notes on the Mammals of the North Bull, Dublin Bay. *Ir. Nat. J.* **3**, 199–201.

O'Mahony, E. (1935). Discovery of a second race of the house mouse in Ireland. *Ir. Nat. J.* **5**, 218–219.

O'Mahony, E. (1935). The North Bull house mouse. *Ir. Nat. J.* **5**, 291.

O'Mahony, E. (1937). On some forms of the house mouse *Mus musculus* Linn. in Ireland. *Ir. Nat. J.* **6**, 288–290.

O'Mahony, E. (1947). The mammal ectoparasites of the North Bull Island, Dublin Bay. *Ir. Nat. J.* **9**, 78–79.

O'Mahony, E. (1949). The fleas of the North Bull Island, Dublin Bay. *Entomologists's mon. Mag.* **85**, 139–140.

O'Reilly, H. & Pantin, G. (1957). Some observations on the salt marsh formation in Co. Dublin. *Proc. R. Ir. Acad.* **58B**, 89–126.

O'Riordan, C. E. (1971). Meiobenthic harpacticoida on the East Coast of Ireland. *Proc. R. Ir. Acad.* **71B**, 191–210.

Pantin, G. (1954). *A study in Irish salt marshes.* Unpublished project report, Department of Botany, University College, Dublin.

Prater, A. J. (1973). *Birds of Estuaries Enquiry* 1971/72. British Trust for Ornithology, Tring.

Prater, A. J. (1974). *Birds of Estuaries Enquiry* 1972/73. British Trust for Ornithology. Royal Society for the Protection of Birds Wildfowl Trust, Tring.

Roden, C. (1973). *An investigation into ecotypic variation in Festuca rubra L. on the North Bull Island, Co. Dublin.* Unpublished project report, Department of Botany, University College, Dublin.

Ruttledge, R. F. (1966). *Ireland's Birds.* Witherby, London.

Ryan, J. & Hayden, T. (1971). *The feeding habits of some birds of the North Bull Island.* Unpublished B.Sc. thesis, University College, Dublin.

Sheehy-Skeffington, M. J. (1976). *Nitrogen fixation and water relations of a dune lichen Peltigera polydactyla (Neck) Hoffm.* Unpublished Moderatorship thesis, Department of Botany, University of Dublin, Trinity College.

Stelfox, A. J. (1951). Description of two small black species of *Dacnusa* (Hym. Braconidae) new to science. *Entomologist's mon. Mag.* **87**, 320–321.

Stronach, B. (1973). The status of Irish wetlands – the current position. In: O'Gorman and Wymes (1973).

Thompson, W. (1849–1851). *The Natural History of Ireland.* Vols. I –III. Reeves, Benham & Reeves, London.

Toner, P. (1967). An Ecological Study of the Genus *Spirorbis* (Serpulidae) on the East Coast of Ireland. *Sci. Proc. R. Dubl. Soc.* **A2**, 319–330.

Ussher, R. J. & Warren, R. (1900). *The Birds of Ireland.* Gurney & Jackson, London.

Watters, J. J. (1853). *The Natural History of the Birds of Ireland.* Dublin & London.

White, J. F. (1966). *A comparison of some methods of vegetation analysis.* Unpublished project report. Department of Botany, University College, Dublin.

ADDITIONAL BIBLIOGRAPHY

Bracken, J. and Kennedy, M. (1967)
Notes on some Irish estuarine and inshore fishes. *Ir. Fish Invest.* Ser. B, *3*, 1–28.

Carpenter, G. H. (1894)
Pycnogonida. *Pallene brevirostris* Johns, and *Ammothea echinata* Hodge in Dublin Bay. *Ir. Nat.*, *3*, 66.

Classens, A. J. M. & O'Rourke, F. J. (1966)
The distribution and general ecology of the Irish Siphonaptera. *Proc. R. Ir. Acad.*, *64*, B, (23), 413–63.

Chandler, P. J. (1975)
An account of the Irish species of two-winged flies (Diptera) belonging to the families of larger Brachycera (Tabanoidea & Asiloidea). *Proc. R. Ir. Acad.*, *75*, B, (2), 81–110.

Goodman, C. (1976)
A survey of the distribution of intertidal lamellibranchs in Dublin Bay in relation to certain physical and chemical parameters, Unpublished B.Sc. Thesis. Department of Zoology, University College, Dublin.

Griffiths, G. C. D. (1968)
Agromyzidae from Ireland. *Proc. R. Ir. Acad.*, *67*, B, (2), 37–61.

Halbert, J. N. (1935)
A revised list of the Irish Hemiptera (Heteroptera and Cicadina). *Proc. R. Ir. Acad.*, *42*, B, (8), 211–318.
1937 Records of Irish Coleoptera. *Ent. mon. Mag.*, *73*, 82–87.

Johnson, W. F. & Halbert, J. N. (1902)
A list of the beetles of Ireland. *Proc. R. Ir. Acad.*, 3rd. series, 6, (4), 535–827.

Nash, R. & Chandler, P. J. (in press)
The Irish species of two-winged flies (Diptera) belonging to the families Fanniidae and Muscidae. *Proc. R. Ir. Acad.*

Praeger, R. L. (ed.) (1929)
Report on recent additions to the Irish fauna and flora (terrestrial and freshwater). *Proc. R. Ir. Acad., 39*, B, (1), 1–94.

Pack-Beresford, D. R. (1908)
Eluma purpurascens, a woodlouse new to the British Isles. *Ir. Nat., 17*, 255–258.
(1909) A supplementary list of the spiders of Ireland. *Proc. R. Ir. Acad., 27*, B, (7), 87–118.

Stelfox, A. W. (1927)
A list of the Hymenoptera Aculeata of Ireland. *Proc. R. Ir. Acad., 37*, B, (22), 201–355.

Speight, M. C. D., Chandler, P. J. & Nash, R. (1975)
Irish Syrphidae (Diptera): notes on the species and an account of their known distribution. *Proc. R. Ir. Acad., 75*, B, (1), 1–80.

Speight, M. C. D. (1976)
Sphaerophoria rueppellii (Dipt. Syrphidae) new to Ireland with figures of its surstyli. *Ir. nat. J., 18*, 332–4.

GLOSSARY

The glossary is designed to assist first reading of chapters without continuous use of several reference books. The entries are defined in the narrow context of particular chapters and more comprehensive definitions will generally be found in the literature.

aerenchyma — Tissue in plants permitting free diffusion of air.

anaerobic — Without oxygen, thus excluding air-requiring organisms.

bar — Unit of pressure, approximately equivalent to 1 atmosphere and conveniently used to express *water potential.*

berm — Convex break of slope on a beach.

biomass — The weight or mass of living organisms in an ecosystem usually expressed in kg/ha.

byssus — Thread-like secretion which anchors mussels to rocks.

chloroplast — The green chlorophyll – containing organelle of plants.

clast — Any sedimented particle.

coelomic fluid — Fluid in the body cavity of invertebrates.

commissure — The line of fusion between two parts of a shelled animal.

consocies — A definable association of plant species, with one species dominant.

debris — Sedimented organic material of large particle size e.g. in tide lines.

detritus — Sedimentary organic particles generally smaller than 1 mm.

dilatant — Firm textured and hardening when pressure is applied, thus difficult to penetrate.

epigenetic variation — Variation due to developmental rather than genetic differences.

euryhaline — Able to tolerate wide variation in environmental salinity.

eutrophication — Process of adding substances, notably phosphorus and nitrogen compounds to an environment in which they are normally in short supply, hence causing biological changes.

Flandrian — Most recent 10,000 years of geological time extending from the close of the Pleistocene.

haustorium — Specialised organ of hemi-parasitic or parasitic plant withdrawing substances from the host.

in-fauna — Animals living beneath the surface of a sediment.

interstitial — Between the grains of sediment.

ion — A charged chemical particle, e.g. when sodium chloride disolves in water it is dissociated into the positively charged cation Na^+ and the anion Cl^-.

lag gravel — Gravel from which finer sediment has been winnowed.

matric potential — See *water potential.*

morainic — Coarse material originally accumulated by the action of a *Pleistocene* ice-sheet.

osmotic potential — See *water potential.*

panicle — A branched racemose inflorescence characteristic of many grasses.

Pleistocene — The last 'Ice-age'. A period extending from 1·5 million years ago to the *Flandrian.*

polyploid — A genetic variant having three or more times the haploid number of chromosomes.

ppm — Parts per million; a unit of concentration expressed as mg/l for solutions or mg/kg for solids.

pyrenoid — A storage granule in the chloroplasts of algae, containing starch and staining black with iodine.

pyritisation — Deposition of iron sulphide as a process in fossilisation.

raptor — A bird of prey often existing at the highest levels of predation in an ecosystem.

reducing conditions Anaerobic condition in which excess of electrons lead to chemical changes such as conversion of Ferric (Fe^{+++}) iron to Ferrous (Fe^{++}).

spat The larvel forms of shellfish.

sporeling A newly germinated plant of a multicellular alga.

sublittoral The lowest zone of a shore, extending from about LWMT to just below LWST.

taxonomy The science of classifying organisms, especially in assigning types to genera and species.

thallus The simple vegetative body of plants such as Algae.

thixotropic Describing sediment which softens and flows if pressure is applied.

tide marks Easily recognisable tidal levels abbreviated as HWST, high water spring tide; HWMT, high water mean tide; LWMT, low water mean tide; LWST, low water spring tide.

unimodal A sediment from a single source.

water potential Usually measured in *bars*. Defines the free energy status of water. Lowering of water potential occurs when substances are dissolved (osmotic potential) or when a matrix, such as soil becomes, unsaturated (matric potential). Water can only move freely from regions of high water potential to regions of low water potential.

VASCULAR PLANTS

Compiled by P. H. Carvill *Department of Botany* *Trinity College*

Over the years botanists have walked the Bull Island, often noting unusual plants. It has always been a happy hunting-ground for the members of the Dublin Naturalists' Field Club and very many of the more than 300 records of vascular plants recorded below come from them, especially from the late J. P. Brunker and from H. J. Hudson. It is hoped that the publication of this list will encourage people rapidly to render it out of date by finding many new records.

Nomenclature follows Scannell & Synnott's (1972) Census Catalogue of the Flora of Ireland. Two codes follow the species name; the first gives information on the major habitat of the species as follows:—

D—Dunes (exclusive of Golf Links and Alder Marsh)

L—Links (including Golf Club premises and the old coast-guard buildings)

M—Alder Marsh

S—Salt-marsh (including the mud flats)

C—New Causeway or disturbed ground associated with it.

The second code letter gives an indication of the abundance and status of the plant as follows:—

A—Plants widely distributed in suitable habitats

B—Plants occurring in a few habitats

C—Plants rare and possibly either recent arrivals or on the verge of extinction

D—Casuals: alien plants, generally of open habitats, which do not persist for more than a few years after introduction; other plants which, although recorded in the past, have not been seen recently

E—Plants which are nationallyrare, or endangered, and which are especially worthy of habitat protection.

Acknowledgements

I would like to thank Miss M. Scannell who advised me on likely sources of additional records and contributed records from the Herbarium. Mr. H. J. Hudson kindly lent me his annotated catalogue and gave me much information on the changing status of plants on the Island. Mr. J. E. Fitzharris provided several additions to Brunker's list and details of some of the rarer plants. Dr. P. H. Pitkin provided records of *Salicornia* and Dr. D. W. Jeffrey advised me on the compilation of this list. I would also like to express my thanks to the Committee of the Dublin Naturalists' Field Club for their assistance in obtaining unpublished records.

PTERIDOPHYTA

EQUISETACEAE

Equisetum variegatum	C	B
Equisetum fluviatile	L	B
Equisetum palustre	M	B
Equisetum arvense	L	B

POLYPODIACEAE

Polypodium vulgare	D	B

OPHIOGLOSSACEAE

Ophioglossum vulgatum	M	C

ANGIOSPERMAE – DICOTYLEDONES

RANUNCULACEAE

Anemone nemorosa	L	C
Clematis vitalba	D	C
Ranunculus acris	D	B
Ranunculus repens	D	B
Ranunculus bulbosus	D	B

PAPAVERACEAE

Papaver rhoeas	L	C
Papaver dubium	L	D

FUMARIACEAE

Fumaria officinalis	ML	B
Fumaria bastardii	L	D

CRUCIFERAE

Brassica rapa	C	D
Brassica nigra	C	C
Sinapis arvensis	CL	A
Diplotaxis muralis	L	C
Cakile maritima	D	B
Lepidium virginicum	D	D
Cardaria draba	L	C
Capsella bursa-pastoris	D	A
Cochlearia officinalis	S	A
Cochlearia anglica	S	A
Erophila verna	D	B
Armoracia rusticana	L	C
Cardamine pratensis	LD	C
Sisymbrium officinale	L	D
Sisymbrium orientale	L	D
Sisymbrium altissimum	L	D

RESEDACEAE

Reseda luteola	C	C

VIOLACEAE

Viola canina	D	A
Viola tricolor ssp. curtisii	D	A

POLYGALACEAE

Polygala vulgaris	D	A

HYPERICACEAE

Hypericum maculatum	L	C
Hypericum perforatum	L	C

CARYOPHYLLACEAE

Silene alba	L	D
Silene vulgaris ssp. maritima	S	B
Silene dichotoma	S	D
Cerastium arvense	L	A
Cerastium fontanum	D	A
Cerastium diffusum	D	A
Cerastium semidecandrum	M	D
Stellaria media	L	B
Sagina maritima	S	A
Sagina nodosa	M	B
Honkenya peploides	D	A
Arenaria serpyllifolia	D	B
Spergula arvensis	D	D
Spergularia rupicola	S	A
Spergularia media	S	A
Spergularia marina	S	D

CHENOPODIACEAE

Chenopodium album	L	A
Chenopodium rubrum	L	D
Beta vulgaris ssp. maritima	S	A
Atriplex littoralis	D	B
Atriplex hastata	SC	C
Atriplex laciniata	D	B
Halimione portulacoides	S	A
Suaeda maritima	S	A
Salsola kali	D	A
Salicornia europaea	S	A
Salicornia dolichostachya	S	A
Salicornia pusilla	S	B

MALVACEAE

Malva moschata	M	D
Malva sylvestris	LC	B
Malva neglecta	L	C

LINACEAE

Linum bienne	D	C
Linum catharticum	D	A

GERANIACEAE

Geranium pyrenaicum	L	C
Geranium dissectum	C	C
Geranium molle	L	A
Geranium robertianum	C	C
Erodium cicutarium	DL	A

ACERACEAE

Acer pseudoplatanus	L	D

HIPPOCASTANACEAE

Aesculus hippocastanum	D	D

PAPILIONACEAE

Ulex europaeus	L	A
Ononis repens	D	A
Medicago sativa	D	C
Medicago lupulina	D	A
Melilotus officinalis	DL	B
Trifolium micranthum	L	B
Trifolium dubium	D	A
Trifolium repens	CL	A
Trifoluim campestre	D	A
Trifolium fragiferum	L	C
Trifolium pratense	DL	A
Trifolium arvense	D	A
Trifolium scabrum	D	CE
Anthyllis vulneraria	D	A
Lotus corniculatus	D	A
Vicia cracca	D	B
Vicia sepium	CD	C
Vicia sativa ssp. nigra (V. angustifolia)	D	A
Lathyrus pratensis	DL	B

ROSACEAE

Filipendula ulmaria	M	B
Rubus fruticosus agg.	L	B
Potentilla anserina	S	B
Potentilla erecta	DL	A
Potentilla reptans	DCL	C
Rosa pimpinellifolia	D	B
Rosa canina	LM	C
Crataegus monogyna	ML	C

CRASSULACEAE

Sedum acre	D	A

ELAEAGNACEAE

Hippophae rhamnoides	L	C

ONAGRACEAE

Epilobium hirsutum	C	C
Epilobium parviflorum	M	C
Epilobium palustre	M	C
Epilobium angustifolium	DL	C
Oenothera stricta	L	C

HYDROCOTYLACEAE

Hydrocotyle vulgaris	M	C

UMBELLIFERAE		
Eryngium maritimum	D	B
Smyrnium olusatrum	L	B
Conopodium majus	L	B
Pimpinella saxifraga	D	B
Aethusa cynapium	L	D
Angelica sylvestris	M	C
Heracleum sphondylium	L	B
Daucus carota	D	A
EUPHORBIACEAE		
Mercurialis annua	LC	C
Euphorbia helioscopia	C	C
Euphorbia peplus	L	D
Euphorbia portlandica	D	B
Euphorbia paralias	D	A
POLYGONACEAE		
Polygonum oxyspermum ssp. raii	D	C
Polygonum aviculare	L	C
Polygonum persicaria	L	C
Reynoutria sachalinensis	L	D
Bilderdykia convolvulus	C	C
Rumex acetosella	D	A
Rumex acetosa	D	C
Rumex crispus	LD	B
Rumex obtusifolius	L	C
Rumex conglomeratus	L	C
URTICACEAE		
Urtica urens	L	C
Urtica dioica	L	C
BETULACEAE		
Alnus glutinosa	M	B
CORYLACEAE		
Corylus avellana	M	C
SALICACEAE		
Salix alba	L	C
Salix atrocinerea	M	C
Salix aurita	M	C
Salix repens	D	C
ERICACEAE		
Erica cinerea	L	C
PLUMBAGINACEAE		
Limonium humile	S	A
Limonium binervosum	S	A
Armeria maritima	S	A
PRIMULACEAE		
Primula veris	L	C
Anagallis tenella	L	C
Anagallis arvensis	D	B
Glaux maritima	S	A
Samolus valerandi	S	B
OLEACEAE		
Fraxinus excelsior	D	D
GENTIANACEAE		
Centaurium pulchellum	SD	BE
Centaurium erythraea	D	A
Blackstonia perfoliata	D	A
Gentianella amarella	M	B
BORAGINACEAE		
Echium vulgare	D	D
Myosotis arvensis	D	B
Myosotis ramosissima	D	C
CONVOLVULACEAE		
Calystegia soldanella	D	B
SOLANACEAE		
Lycium chinense	LM	C
Solanum dulcamara	D	C
Solanum nigrum		D
Solanum tuberosum		D
SCROPHULARIACEAE		
Verbascum thapsus	L	C
Veronica hederifolia	L	C
Veronica persica	L	C
Pedicularis sylvatica	D	A
Rhinanthus minor	D	A
Euphrasia officinalis agg.	D	A
Euphrasia nemorosa	D	
Euphrasia arctica ssp. borealis	D	D
Odontites verna	DC	B
OROBANCHACEAE		
Orobanche minor	DC	B
LENTIBULARIACEAE		
Pinguicula vulgaris	M	C
LABIATAE		
Mentha aquatica	M	B
Thymus drucei	C	C
Lycopus europaeus	M	C
Salvia horminoides	L	B
Prunella vulgaris	D	A
Stachys palustris	C	C
Stachys sylvatica	C	C
Lamium hybridum	C	C
Lamium purpureum	L	C
Galeopsis angustifolia	CL	C
PLANTAGINACEAE		
Plantago major	L	C
Plantago media	L	D
Plantago lanceolata	L	A
Plantago maritima	S	A
Plantago coronopus	S	A
CAMPANULACEAE		
Campanula persicifolia	L	C
Campanula rotundifolia	L	C
RUBIACEAE		
Sherardia arvensis	L	C
Galium verum	D	B
Galium palustre	M	C
Galium aparine	L	C
CAPRIFOLIACEAE		
Sambucus nigra	M	B
VALERIANACEAE		
Valerianella locusta	D	D
Valeriana officinalis	LM	C
DISPSACACEAE		
Succisa pratensis	M	C
COMPOSITAE		
Senecio jacobaea	D	A
Senecio vulgaris	D	A
Senecio cineraria	C	C
Senecio erucifolius	LM	C
Senecio squalidus	C	C
Tussilago farfara	C	C
Petasites fragrans	C	C
Pulicaria dysenterica	M	C
Antennaria dioica	L	C
Filago germanica	L	D
Aster tripolium	S	A
Erigeron acer	L	C
Bellis perennis	L	A
Achillea millefolium	L	C
Tripleurospermum maritimum ssp. maritimum	S	A
Chrysanthemum segetum	L	C
Leucanthemum vulgare	D	A
Tanacetum parthenium	C	C
Matricaria matricarioides	L	B
Artemisia vulgaris	L	C

Artemisia stellerana	Extinct	
Carlina vulgaris	D	A
Arctium lappa	C	C
Cirsium vulgare	L	C
Cirsium palustre	L	C
Cirsium arvense	D	C
Silybum marianum	L	D
Centaurea nigra	D	B
Hypochoeris radicata	D	A
Lapsana communis	C	C
Leontodon hispidus	D	B
Leontodon autumnalis	S	B
Leontodon taraxacoides	D	B
Tragopogon pratensis	L	C
Sonchus arvensis	L	B
Sonchus oleraceus	L	B
Sonchus asper	D	C
Pilosella officinarum	LD	B
Crepis taraxacifolia	D	C
Crepis capillaris	D	B
Taraxacum officinale	D	A
Taraxacum palustre	M	C
MONOCOTYLEDONES		
JUNCAGINACEAE		
Triglochin palustris	L	C
Triglochin maritima	S	A
POTAMOGETONACEAE		
Potamogeton pusillus	S	C
Ruppia maritima	S	B
Zostera marina	S	C
Zostera noltii	S	B
Zostera angustifolia	S	B
LILIACEAE		
Asparagus officinalis	D	B
Ornithogalum pyrenaicum	L	C
Scilla non-scripta	L	C
Allium vineale	L	B
JUNCACEAE		
Juncus gerardii	S	A
Juncus bufonius	L	B
Juncus ambiguus Guss.	S	—
Juncus maritimus	S	A
Juncus acutus	LS	B
Luzula campestris	D	A
AMARYLLIDACEAE		
Narcissus pseudonarcissus	L	D
IRIDACEAE		
Iris psuedacorus	D	C
Crocus nudiflorus	D	D
Crocosmia x crocosmiflora	D	C
ORCHIDACEAE		
Epipactis palustris	ML	B
Spiranthes spiralis	L	C
Listera ovata	M	C
Gymnadenia conopsea	D	C
Ophrys apifera	D	B
Orchis morio	L	C
Dactylorhiza incarnata	D	B
Dactylorhiza maculata	L	C
Dactylorhiza fuchsii	D	B
Dactylorhiza purpurella	M	C
Anacamptis pyramidalis	D	A
CYPERACEAE		
Eriophorum angustifolium	L	C
Eleocharis quinqueflora	L	C
Blysmus rufus	S	B
Carex serotina	D	A
Carex extensa	D	A
Carex flacca	D	C
Carex caryophyllea	D	A
Carex nigra	M	C
Carex otrubae	M	C
Carex arenaria	D	A
Carex ovalis	M	C
Carex pulicaris	L	C
GRAMINEAE		
Festuca rubra	DS	A
Festuca ovina	C	C
x Festulolium loliaceum		C
Lolium perenne	DL	B
Lolium multiflorum	L	C
Puccinellia maritima	S	A
Catapodium rigidum	L	C
Catapodium marinum	D	B
Poa annua	D	C
Poa pratensis	D	B
Dactylis glomerata	D	C
Cynosurus cristatus	DL	B
Briza media	D	A
Bromus mollis	D	B
Agropyron junceiforme	D	A
Elymus arenarius	D	B
Hordeum murinum	L	C
Koeleria cristata	D	C
Trisetum flavescens	D	B
Avena fatua	C	C
Arrhenatherum elatius	D	B
Helictotrichon pubescens	D	B
Holcus lanatus	D	B
Aira caryophyllea	D	B
Aira praecox	L	B
Ammophila arenaria	D	A
Agrostis tenuis		C
Agrostis stolonifera	S	C
Polypogon monspeliensis	L	D
Phleum pratense	C	C
Phleum arenarium	D	D
Anthoxanthum odoratum	D	B
Phalaris canariensis	C	C
Parapholis strigosa		C
Spartina x townsendii	S	B
Spartina anglica	S	A
Spartina maritima forma dublinensis	S	C

MOSSES AND LIVERWORTS

P. H. Pitkin *Department of Botany* *Trinity College*

Arrangement and nomenclature follow Warburg (1965) for mosses and Paton (1965) for liverworts. All records were checked in the field during 1974–1975 by the author, and only seven records are classified as either doubtful or possibly now extinct, the species not being refound. *Bryum bicolor* var* is a variety with several small bulbils in each leaf axil. It was first found at Cassington in Oxfordshire in 1948 and has yet to be named. It is known from a few damp sandy places in Co. Wicklow and Co. Wexford.

Abundance Symbols

A	Common in suitable habitats
B	Occasional or local
C	Rare
D	Plants not seen recently

Habitat Symbols

1 Mobile sand.
2 Fixed sand with grassy turf.
3 Fixed sand with no turf, stabilised by mosses and annuals.
4 Old sand dunes, e.g. between fairways on St. Anne's Golf Course.
5 Damp sand with dense cover of higher plants, e.g. the Alder Marsh.
6 Damp sand with sparse cover of higher plants, e.g. the group of slacks near the fence at S.W. end of Royal Dublin Golf Course.
7 Damp sand bare of higher plants, e.g. behind the fore-dunes at S. end of the beach.
8 Disturbed ground on the golf course, usually where turf has been removed.
9 On built-up land.

MUSCI

Polytrichum juniperinum	B 4
Fissidens adianthoides	B5
Ditrichum flexicaule	D
Ceratodon purpureus	A 1 3 9
Dicranella varia	B 8
Dicranum scoparium	B 4
Tortula ruralis	B 3
T. ruraliformis	A 1 3
T. subulata	B 3
Pottia heimii	C 8
Phascum cuspidatum	C 8
Barbula convoluta	A 6 9
B. unguiculata	B 3 6 9
B. hornschuchiana	C 6
B. fallax	A 3 6 9
B. reflexa	B 6
B. rigidula	B 3
B. tophacea	A 6 7
B. cylindrica	A 6
B. recurvirostra	B 6
Trichostomum brachydontium	C 6
Tortella flavovirens	D
Funaria hygrometrica	B 7 8 9
Leptobryum piriforme	C 8
Pohlia annotina	C 8
Pohlia delicatula	C 6 8
Bryum marratii	C 8
B. warneum	B 6 7
B. pendulum	A 1 3 5 6 7 8
B. inclinatum	A 5 6 7 8
B. calophyllum	D
B. pseudotriquetrum	A 5 6
B. intermedium	D
B. caespiticium	C 3
B. argentium	B 3 7 8 9
B. argentium var lanatum	C 3
B. bicolor	A 6 7 8 9
B. bicolor var*	7 8
Mnium undulatum	C 2
Amblyodon dealbatus	C 6
Climacium dendroides	C 5
Thuidium tamariscinum	B 4
Cratoneuron filicinum	B 6
Campylium stellatum	B 5 6
Leptodictyum riparium	C 8
Amblystegium serpens var. salinum	B 6
Drepanocladus aduncus	C 6
Acrocladium cuspidatum	A 5 6
Camptothecium lutescens	B 2 3
Brachythecium albicans	A 1 2 3
B. mildeanum	B 2 6
B. rutatulum	B 2 5 6
Eurhynchium praelongum	B 2 5 6
E. megapolitanum	B 2
Pseudoscleropodium purum	B 2 4
Hypnum cupressiforme var. tectorum	A 2 3 4
H. cupressiforme var. lacunosum	B 2
Rhytidiadelphus triquetrus	B 2 4
R. squarrosus	B 2 4
Hylocomium splendens	B 4

HEPATICAE

Reboulia hemisphaerica	D
Preissia quadrata	D
Riccardia sinuata	B 5 6
R. pinguis	A 5 6
Pellia indiviifolia	B 5 6
Moerekia flotoviana	D
Petalophyllum ralfsii	C 6
Lophocolea bidentata	B 2 5 6

ALGAE

P. H. Pitkin *Department of Botany* *Trinity College*

The following list includes only the multicellular and filamentous Rhodophyta, Phaeophyta, Chlorophyta and Xanthophyta found on the intertidal flats. The order and nomenclature follow Parke & Dixon (1968).

Rhodophyta
Chondrus crispus
Antithamnium cruciatum
Ceramium rubrum
C. tenuissimum
Polysiphonia macrocarpa
Porphyra umbilicalis

Xanthophyta
Vaucheria velutina

Phaeophyta
Ectocarpus silicuosus
Pilayella littoralis
Dictyosiphon foeniculaceus
Scytosiphon lomentarius

Sphacelaria pumila
Ascophyllum nodosum
Fucus ceranoides
F. serratus
F. spiralis
F. vesiculosus
Pelvetia canaliculata

Chlorophyta
Gomontia polyrhiza
Ostreobium queckettii
Monostroma grevilliae
Blidingia marginata
B. minima
Enteromorpha compressa
Enteromorpha flexuosa
E. intestinalis
E. linza
E. prolifera
E. ralfsii
E. ramulosa
E. torta
E. usneoides
Percursaria percursa
Ulva lactuca
Chaetomorpha linum
Cladophora fracta
Rhizoclonium arenosum
R. implexum

LICHENS

Compiled by C. C. Moore

Department of Botany, University College, Dublin

Twenty sites were examined for lichens in Spring, 1976 *(Fig. A1)*. The species found are listed below with the genera arranged in alphabetical order.
The nomenclature follows P. W. James' A new checklist of British Lichens (1965). The habitats are indicated by the following code:

W—Confined to the Bull Wall (i.e. the area lying between the statue of Mary and a point opposite the Royal Dublin Clubhouse); principally limestone rocks lying between low water and the road surface.

T—Trees in Alder Marsh.

M—Ground in Alder Marsh.

D—Dunes.

S—Saxicolous substrates generally (including the Bull Wall).

V—Corticolous and saxicolous substrates (lignicolous in the case of *Lecanora conizaeoides)*.

X—Most terricolous habitats except the links and younger dunes.

The scale used to indicate degrees of abundance (2nd column) is as follows:—

A—Very common in suitable habitat

B—Common in suitable habitat

C—Scarce

D—Rare

E—Scarce to rare and depauperate due to air pollution.

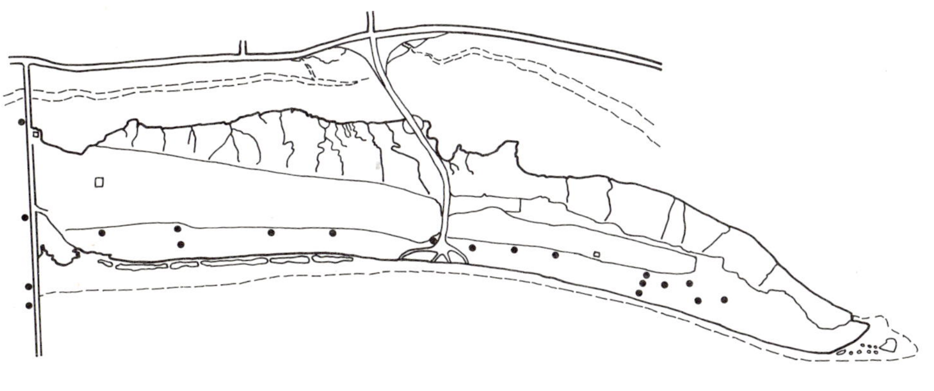

Acarospora fuscata	W	D
Anaptychia fusca	W	D
Arthopyrenia halodytes	W	B
Bacidia muscorum	M	D
Buellia canescens	W	C
Buellia punctata	T	B
Buellia subdisciformis	W	C
Caloplaca citrina	W	C
Caloplaca marina	W	B
Caloplaca murorum	S	C
Caloplaca thallincola	W	B
Candelariella aurella	S	B
Candelariella vitellina	W	B
Cladonia rangiformis	D	A
Cladonia impexa	D	B
Cladonia foliacea	D	D
Cladonia squamosa	M	C
Cladonia coniocraea	T	D
Cladonia fimbriata	D	B
Cladonia pyxidata	M	C
Collema tenax var. ceranoides	M	B
Collema limosum	M	C
Diploschistes scruposus var. bryophilus	M	C
Evernia prunastri	T	E
Lecania erysibe	W	C
Lecanora atra	S	B
Lecanora actophila	W	C
Lecanora gangaleoides	S	A

Lecanora helicopis	W	B	Parmelia physodes	T	E	Rinodina subexigua	W	C
Lecanora calcarea	W	C	Parmelia sulcata	T	E	Sarcogyne simplex	W	D
Lecanora chlarotera	T	D	Parmelia glabratula	T	E	Verrucaria maura	W	B
Lecanora expallens	T	C	Peltigera canina	X	A	Verrucaria striatula	W	B
Lecanora conizaeoides	V	B	Peltigera polydactyla	M	C	Verrucaria mucosa	W	C
Lecanora dispersa	S	B	Peltigera rufescens	M	D	Verrucaria microspora	W	C
Lecidea stigmatea	S	B	Physcia adscendens	T	E	Verrucaria muralis	S	C
Lepraria incana	T	C	Physcia tenella	W	C	Verrucaria nigrescens	S	C
Leptogium lichenoides	M	B	Physcia orbicularis	W	C	Xanthoria parietina	V	B
Lichina pygmaea	W	C	Physcia pulverulenta	T	D	Xanthoria polycarpa	T	D
Lichina confinis	W	C	Ramalina farinacea	T	E			
Opegrapha calcarea	W	C	Rhizocarpon constrictum	W	D			

THE FUNGI

M. J. P. Scannell *The Herbarium National Botanic Garden Glasnevin*

The fungus flora of the North Bull with 200 recorded species is rich despite the limited range of habitat and soil types. The area has not been, as yet, fully or systematically worked. Present knowledge of the fungus flora is due to two mycologists, Dr. Patrick O'Connor (1889–1969) for many years Keeper of Natural History, National Museum of Ireland, and Mr. F. C. Hassel, (fl. 1948–1956), officer of the Revenue Commissioners and an amateur botanist in his leisure hours. Both were members of the Dublin Naturalists Field Club.

O'Connor studied primarily the stem and leaf fungi of plants in the wild. In the course of field work throughout Ireland he collected many specimens of saprophytes and parasites many new to the Flora of Ireland. He added greatly to our knowledge of rust and smut fungi of native plants. His collection of documented specimens forms the core of the Irish Mycological collections in the Herbarium, (DBN) National Botanic Gardens, Glasnevin. He published his work in the Royal Dublin Society Proceedings.

Hassel studied mainly the macrofungi, Basidiomycetes (mushrooms) and Ascomycetes (cup fungi). He confined his field work to the North Bull, as it was convenient to the city of Dublin. In 1953 he published in the Irish Naturalists Journal a list of species of fungi, a contribution towards a survey conducted by the Dublin Naturalists Field Club. Some species had not been previously recorded in the Irish flora. The 1953 list provides the bulk of the species in the present list. Some of the specimens are represented in the National Herbarium.

Subsequent to the work of O'Connor and Hassel little mycological activity has taken place in the North Bull. Some specimens collected by M. Scannell have been included in the list below.

Status of Nomenclature

All names in the following list have been checked for modern revisions in Dennis (1960) (D) or Singer (1961) (S). Where types were not listed by these authorities Rea (1922) and Grove (1935) were used and (R) or (G) respectively, appear after the name. Where the name could not be found in any available taxonomic work (?) follows the name. Taxonomic revision by P. Dowding, Botany Department, Trinity College.

Habitats

Habitats, substrates or associated plants are abbreviated as follows:—D – Dunes, L – Links, M – Alder Marsh, F – Dung, S – *Salix* patches, Mo – Moss, Gr – Grass, Am – *Ammophila*, Al – *Alnus*, 1 – *Populus nigra*, 2 – *Aster tripolium*, 3 – *Carex extensa*, 4 – *Allium sp.*, 5 – *Lotus corniculatus*, 6 – *Anthyllis vulneraria*, 7 – *Agropyron junceiforme*, 8 – *Filipendula ulmaria*, 9 – *Angelica (Smyrnium olusatrum?)* 10 – *Rumex sp.*, 11 – *Lycium halimifolium*, 12 – *Cirsium arvense*, 13 – *Salix repens*, 14 – *Juncus sp.*, 15 – *Carex sp.*, 16 – *Sisymbrium sp.*, 17 – *Potato*, 18 – *Senecio jacobea*, 19 – *Malva sylvestris*, 20 – *Trifolium repens*, 21 – *Limonium humile*, 22 – *Salicornia sp.*, ? – No habitat record.

Fungi Imperfecti

SPHAEROPSIDALES

Amerosporium caricum	(G)	15
Ascochyta stipae f. agropyri	(G)	7
Cercospora beticola	(G)	?
Cicinobolus cesata	(G)	?
Cytospora salicis	(G)	13
Leptostroma donacinium	(G)	Am
Leptothyrium asterinum	(G)	2
Phoma acuta	(G)	16
Phyllosticta statices	(G)	21
Stagonospora compta	(G)	20
Stagonosporopsis salicorniae	(G)	22

MONILIALES

Alternaria solani	(?)	17
Botrytis cinerea	(?)	17
Capularia arundini	(?)	Am
Cladosporium herbarum	(?)	Am
Dematium hispidulum	(?)	Am
Epicoccum purpurascens	(?)	Am
Fusarium sp.	—	18
Heterosporium allii	(?)	4
Periconia byssoides	(?)	19
P. igniaria	(?)	Am
Polythrincium trifolii	(?)	20
Tetraploa aristata	(?)	Gr
Torula herbarum	(?)	19
Trichothecium roseum	(?)	18
Trichoderma viride	(?)	Ul

Basidiomycetes

GASTEROMYCETALES

Bovista nigrescens	(R)	D, L
B. plumbea	(R)	L
Cyathus olla	(R)	D
Lycoperdon depressum	(R)	L
L. saccatum	(R)	L
L. spadiceum	(R)	D
L. umbrinum	(R)	D
Tulostoma brumale	(R)	D

AGARICALES

Agaricus augustus	(S)	L
A. bisporus	(S)	D
A. campestris	(S)	L
A. comtulus	(S)	L
A. silvicola	(S)	D
Agrocybe dura	(S)	D
A. praecox	(S)	M
A. praecox var paludosa	(S)	M
A. pumila	(S)	M
A. semiorbicularis	(S)	D
A. vervacti	(S)	D
Alnicola melinoides	(S)	Al
A. sclolecina	(S)	Al
Bolbitius vitellinus	(S)	L
Calocybe carnea	(S)	D
C. gambosa	(S)	L
C. ionides	(S)	D
Camarophyllus niveus	(S)	L
C. virgineus	(S)	L
Cantharellus copulatus	(?)	D
Clitocybe dealbata	(S)	D
C. ditopa	(S)	D
C. luffi	(S)	D
C. obsoleta	(S)	D
C. rivulosa	(S)	D
C. umbilicata	(S)	D
Conocybe lactea	(S)	L
C. tenera	(S)	L
Coprinus cinereus	(S)	M
C. cordisporus	(R)	F
C. macrocephalus	(R)	F
C. plicatilis	(S)	L
Cortinarius ammophilus	(S)	S
C. paleaceus	(S)	S
Crinipellis stipitaria	(S)	D
Cystoderma amianthinum	(S)	L
Flammulina velutipes	(S)	Al
Galerina badipes	(S)	L
G. clarata	(S)	L
G. graminea	(S)	L, D
G. hypnorum	(S)	L
G. mniophila	(S)	L
G. mniophila	(S)	L
Hebeloma crustaliniforme	(S)	S
H. mesophaeum	(S)	S
Hemimycena mairei	(S)	D
Hohenbuehelia geogenia	(S)	
H. petaloides	(S)	?
Hygrocybe autocarnea	(S)	L
H. coccinea	(S)	L
H. conica	(S)	D
H. nigrescens	(S)	D
H. obrussea	(S)	L
H. psittacina	(S)	L
H. punicea	(S)	D
Hygrophorus russo-coriaceus	(?)	L
Inocybe brumea	(S)	D
I. serolina	(S)	S
Lepiota alba	(R)	D
Lepista lascina	(S)	L
L. nuda	(S)	D
L. personata	(S)	L
L. sordida	(S)	L
Leptoglossum muscigenium	(S)	?
Leucoagaricus naucina	(S)	D
Marasmius littoralis	(S)	D
M. oreades	(S)	D, L
Mycena acicula	(S)	S
M. citrinomarginata	(S)	S
M. chlorantha	(S)	?
M. chlorinella	(S)	L
M. epipterygia	(S)	Al
M. hiemalis	(S)	Al
M. pura	(S)	L
M. quisquiliaris	(S)	M
M. tenerrima	(S)	Al
M. vitilis	(S)	F
Omphalia hepatica	(R)	D
O. leucophylla	(R)	L
Omphalina demissa	(S)	D
O. ericetorum	(S)	D
O. pyxidata	(S)	D
O. rustica	(S)	D
Panaeolus ater	(S)	D
P. fimicola	(S)	D
P. papilionaceus	(S)	D
P. sphinctrinus	(S)	D
Panaeolina foenisecii	(S)	L
Pholiota flavida	(S)	Am
Pleurotellus chioneus	(S)	Gr
Psalliota arenicola	(?)	D
Psathyrella ammophila	(S)	D
P. spadicea	(S)	S
P. subatrata	(S)	Al
Psilocybe coprophila	(S)	D
P. inquilina	(S)	D
P. montana	(S)	L
P. physaloides	(S)	Gr.
Rhodopaxillas fallax	(S)	D
R. popinalis	(S)	D
Rhodophyllus clandestina	(S)	?
R. griseo-cyaneus	(S)	L
R. sericeus	(S)	D
R. undatus	(S)	L
Stropharia aeruginosa	(S)	?
S. albo-cyanea	(S)	L
S. coronilla	(S)	D
S. inuncta	(S)	D
S. semi-globata	(S)	L
Tomentella livida	(?)	S
Tubaria erabola	(R)	L

T. furfuracea	(S)	S
Volvariella speciosa	(S)	L

APHYLLOPHORALES

Clavaria luteo-alba	(R)	L
C. tenuipes	(R)	M
Coniophora sp.	—	Ul
Corticium confluens	(R)	Ul, Al
C. laeve	(R)	S
Grandinia brinkmannii	(R)	Ul
Irpex obliquus	(R)	Ul
Merulius corium	(R)	Ul
Mycoleptodon ochracum	(R)	Ul
Odontia corrugata	(?)	S
Peniophora cinerea	(R)	Ul, Al
P. setifera	(R)	?
P. incarnata	(R)	Ul
Polystictus versicolor	(R)	Ul
Solenia anomala	(R)	S
Stereum hirsutum	(R)	Ul

UREDINALES

Melampsora larici-populina	(?)	1
M. orchidi-repentis	(?)	?
Puccinia extensicola	(?)	2, 3
P. porri	(?)	4
Uromyces loti	(?)	5
U. anthyllidis	(?)	6

TREMELLALES

Dacryomyces deliquescens	(?)	Ul

Ascomycetes

PYRENOMYCETES

Chaetosphaeria myriocarpa	(D)	Ul
Claviceps purpurea	(D)	7
Clypeosphaeria notarisii	(D)	8
Daldinia concentrica	(D)	Ul
Diaporthopsis angelicae	(D)	9
Gaumannomyces rubellus	(D)	10
Gibberella cyanogena	(D)	11
Herpotrichia macrotricha	(D)	S
Hypoxylon fuscum	(D)	Al
Hysterium angustatum	(D)	Al
Leptosphaeria michottii	(?)	Am
Lophiotrema angustilabrum	(D)	Ul
Lophodermum arundinaceum	(D)	Am
Nectria cinnabarina	(D)	11
Pleospora herbarum	(D)	12
Rosellinia mammiformis	(D)	13
Scutellaria trechispora	(D)	Ul

DISCOMYCETES

Belonioscypha vexata	(D)	10
Dasyscyphus carneola	(D)	Am
D. diminutus	(D)	14
D. palearum	(D)	Am
D. pudicella	(D)	Gr
D. virgineus	(D)	Al
Geoglossum cookeianum	(D)	D
Heliotium robustius	(D)	8
H. scutula	(D)	8
Hyalina rubella	(D)	S
Hyaloscypha deparcula	(D)	8
Mollisia discolor var. longispora	(D)	Al
M. ligni	(D)	Ul Al
M. ventosa	(D)	Al
Peziza vesiculosa	(D)	F

SYSTEMATIC LIST OF BIRDS RECORDED ON THE NORTH BULL

R. N. Goodwillie *An Foras Forbartha*

A total of 180 species of birds had been recorded on the North Bull up to 31 October 1975. This compares with the 276 which had been recorded in the two counties of Dublin and Wicklow up to the end of 1973 (Hutchinson, 1975). A few changes in Kennedy's (1953) list have been deemed necessary bringing its total number of forms to 147. The 33 additional forms are indicated with an asterisk in the list that follows. The list is arranged to follow the sequence and nomenclature of Hudson (1971).

The status of each species during the four seasons of the year is summarised using the following symbols:

I – Isolated records since 1953
R – Rare; less than one bird per year
S – Scarce; 1 – 20 birds per year
C – Common; 21 – 500 birds per year
A – Abundant; over 500 birds per year

Black-throated Diver *Gavia arctica* (Linnaeus)	R	—	R	R
Great Northern Diver *Gavia immer* (Brunnich)	S	—	S	S
Red-throated Diver *Gavia stellata* (Pontoppidan)	C	—	C	C
Great Crested Grebe *Podiceps cristatus* (Linnaeus)	S	—	S	S
Red-necked Grebe *Podiceps griseigena* (Boddaert)	R	—	—	R
Slavonian Grebe *Podiceps auritus* (Linnaeus)	I	—	I	I
Black-necked Grebe *Podiceps nigricollis* (Brehm)	—	—	—	I
Little Grebe *Tachybaptus ruficollis* (Pallas)	S	—	S	S
Fulmar *Fulmarus glacialis* (Linnaeus)	S	S	S	R

*Manx Shearwater *Puffinus puffinus* (Brunnich)	—	S	S	—
Gannet *Sula bassana* (Linnaeus)	S	S	S	—
Cormorant *Phalacrocorax carbo* (Linnaeus)	C	C	C	C
Shag *Phalacrocorax aristotelis* (Linnaeus)	C	C	C	C
Grey Heron *Ardea cinerea* (Linnaeus)	C	S	C	C
*Little Egret *Egretta garzetta* (Linnaeus)	I	—	—	—
*Spoonbill *Platalea leucordia* (Linnaeus)	—	—	I	I
Glossy Ibis *Plegadis falcinellus* (Linnaeus)	R	—	R	R
Mallard *Anas platyrhynchos* (Linnaeus)	C	C	C	C
Teal *Anas crecca* (Linnaeus)	C	R	A	A
*Blue-Winged Teal *Anas discors* (Linnaeus)	—	—	I	I
Gadwall *Anas strepera* (Linnaeus)	R	—	R	R
Wigeon *Anas penelope* (Linnaeus)	C	—	A	A
*American Wigeon *Anas americana* (Gmelin)	—	—	—	I
Pintail *Anas acuta* (Linnaeus)	C	—	C	A
Shoveler *Anas clypeata* (Linnaeus)	C	—	C	C
Scaup *Aythya marila* (Linnaeus)	S	—	S	S
Tufted Duck *Aythya fuligula* (Linnaeus)	R	R	R	R
Pochard *Aythya ferina* (Linnaeus)	R	R	R	R
Goldeneye *Bucephala clangula* (Linnaeus)	S	—	S	S
Long-tailed Duck *Clangula hyemalis* (Linnaeus)	S	—	S	S
Velvet Scoter *Melanitta fusca* (Linnaeus)	—	—	R	R
Surf Scoter *Melanitta perspicillata* (Linnaeus)	—	—	I	I
Common Scoter *Melanitta nigra* (Linnaeus)	C	R	C	C
Eider *Somateria mollissima* (Linnaeus)	R	—	R	R
Red-breasted Merganser *Mergus serrator* (Linnaeus)	C	S	C	C
Goosander *Mergus merganser* (Linnaeus)	—	—	—	I
Smew *Mergus albellus* (Linnaeus)	—	—	—	I
Shelduck *Tadorna tadorna* (Linnaeus)	A	S	S	A
Greylag Goose *Anser anser* (Linnaeus)	—	—	R	R
White-fronted Goose *Anser albifrons* (Scopoli)	—	—	I	I
*Pink-footed Goose *Anser brachyrhynchus* (Baillon)	—	—	I	I
Brent Goose *Branta bernicla* (Linnaeus)	A	R	C	A

Barnacle Goose *Branta leucopsis* (Bechstein)	R	—	R	R
Mute Swan *Cygnus olor* (Gmelin)	S	S	S	S
Whooper Swan *Cygnus cygnus* (Linnaeus)	R	—	R	R
Berwick's Swan *Cygnus bewickii* (Yarrell)	R	—	R	R
Buzzard *Buteo buteo* (Linnaeus)	—	—	R	R
Sparrowhawk *Accipiter nisus* (Linnaeus)	S	S	S	S
Hen Harrier *Circus cyaneus* (Linnaeus)	—	—	R	R
Montagu's Harrier *Circus pygargus* (Linnaeus)	R	—	—	—
Peregrine *Falco peregrinus* (Tunstall)	S	R	S	S
Merlin *Falco columbarius* (Linnaeus)	S	—	S	S
Kestrel *Falco tinnunculus* (Linnaeus)	S	—	S	S
Pheasant *Phasianus colchicus* (Linnaeus)	—	—	—	R
Water Rail *Rallus aquaticus* (Linnaeus)	—	—	—	R
Moorhen *Gallinula chloropus* (Linnaeus)	S	S	S	S
Oystercatcher *Haematopus ostralegus* (Linnaeus)	A	A	A	A
Lapwing *Vanellus vanellus* (Linnaeus)	C	C	C	C
Ringed Plover *Charadrius hiaticula* (Linnaeus)	C	C	C	C
*Little Ringed Plover *Charadrius dubius* (Scopoli)	—	—	I	—
Kentish Plover *Charadrius alexandrinus* (Linnaeus)	—	—	R	—
Grey Plover *Pluvialis squatarola* (Linnaeus)	A	S	C	A
Golden Plover *Pluvialis apricaria* (Linnaeus)	A	S	C	A
*Lesser Golden Plover *Pluvialis dominica* (P. L. S. Muller)	—	—	I	—
Turnstone *Arenaria interpres* (Linnaeus)	C	C	C	C
*Dowitcher *Limnodromus sp.* (Wied)	—	—	I	—
Snipe *Gallinago gallinago* (Linnaeus)	C	R	C	C
Great Snipe *Gallinago media*	—	—	—	R
Jack Snipe *Lymnocryptes minimus* (Brunnich)	R	—	R	R
Woodcock *Scolopax rusticola* (Linnaeus)	—	—	—	R
Curlew *Numenius arquata* (Linnaeus)	A	C	A	A
Whimbrel *Numenius phaeopus* (Linnaeus)	C	S	C	R
Black-tailed Godwit *Limosa limosa* (Linnaeus)	S	—	C	C
Bar-tailed Godwit *Limosa lapponica* (Linnaeus)	A	S	A	A

Green Sandpiper *Tringa ochropus* (Linnaeus)	—	—	R	—
*Wood Sandpiper *Tringa glareola* (Linnaeus)	R	—	R	—
Common Sandpiper *Tringa hypoleucos* (Linnaeus)	S	—	S	—
Redshank *Tringa totanus* (Linnaeus)	A	S	A	A
Spotted Redshank *Tringa erythropus* (Pallas)	R	R	S	S
Greenshank *Tringa nebularia* (Gunnerus)	C	R	C	C
Knot *Calidris canurus* (Linnaeus)	A	—	C	A
Purple Sandpiper *Calidris maritima* (Brunnich)	—	—	—	R
Little Stint *Calidris minuta* (Leisler)	R	—	S	—
*Pectoral Sandpiper *Calidris melanotos* (Vieillot)	—	—	I	—
Dunlin *Calidris alpina* (Linnaeus)	A	C	A	A
Curlew Sandpiper *Calidris ferruginea* (Pontopiddan)	—	—	S	—
Sanderling *Calidris alba* (Pallas)	C	—	C	C
*Buff-Brested Sandpiper *Tryngites subruficollis* (Vieillot)	—	—	I	—
Ruff *Philomachus pugnax* (Linnaeus)	R	—	S	—
Avocet *Recurvirostra avosetta* (Linnaeus)	—	—	—	I
Grey Phalarope *Phalaropus fulicarius* (Linnaeus)	—	—	R	—
*Red-necked Phalarope *Phalaropus lobatus* (Linnaeus)	—	—	—	I
*Wilson's Phalarope *Phalaropus tricolor* (Vieillot)	—	—	R	—
Stone Curlew *Burhinus oedicnemus* (Linnaeus)	—	—	—	R
Pomarine Skua *Stercorarius pomarinus* (Temminck)	—	—	—	R
Arctic Skua *Stercorarius parasiticus* (Linnaeus)	—	—	R	—
Great Black-backed Gull *Larus marinus* (Linnaeus)	C	C	C	C
Lesser Black-backed Gull *Larus fuscus* (Linnaeus)	C	C	C	R
Herring Gull *Larus argentatus* (Pontoppidan)	A	A	A	A
Common Gull *Larus canus* (Linnaeus)	A	S	A	A
Iceland Gull *Larus glaucoides* (Meyer)	—	—	—	I
Glaucous Gull *Larus hyperboreus* (Gunnerus)	R	—	R	R
*Little Gull *Larus minutus* (Pallas)	—	—	R	—
Black-headed Gull *Larus ridibundus* (Linnaeus)	A	C	A	A
Sabine's Gull *Larus sabini* (Sabine)	—	—	R	—
Kittiwake *Rissa tridactyla* (Linnaeus)	C	S	C	S
*Black Tern *Chlidonias niger* (Linnaeus)	I	—	I	—
Common Tern *Sterna hirundo* (Linnaeus)	C	C	C	—
Artic Tern *Sterna paradisaea* (Pontoppidan)	C	C	C	—
Roseate Tern *Sterna dougallii* (Montagu)	R	—	C	—
*Bridled Tern *Sterna anaethetus* (Scopoli)	—	—	—	R
Little Tern *Sterna albifrons* (Pallas)	C	C	C	—
*Royal Tern *Sterna maxima* (Boddaert)	R	—	—	—
Sandwich Tern *Sterna sandvicensis* (Latham)	C	C	C	—
Razorbill *Alca torda* (Linnaeus)	S	S	S	S
Little Auk *Plautus alle* (Linnaeus)	—	—	—	R
Guillemot *Uria aalge* (Pontoppidan)	S	S	S	S
Puffin *Fratercula arctica* (Linnaeus)	R	R	—	—
Woodpigeon *Columba palumbus* (Linnaeus)	S	S	S	S
Cuckoo *Cuculus canorus* (Linnaeus)	R	—	—	R
Snowy Owl *Nyctea scandiaca* (Linnaeus)	—	—	—	R
Long-eared Owl *Asio otus* (Linnaeus)	R	S	R	R
Short-eared Owl *Asio flammeus* (Pontoppidan)	S	—	S	S
Nightjar *Caprimulgus europaeus* (Linnaeus)	—	—	R	—
Swift *Apus apus* (Linnaeus)	S	S	S	—
Kingfisher *Alcedo atthis* (Linnaeus)	R	S	S	S
*Wryneck *Jynx torquilla* (Linnaeus)	—	—	I	—
Skylark *Alauda arvensis* (Linnaeus)	C	C	C	C
*Shore Lark *Eremophila alpestris* (Linnaeus)	—	—	R	—
Swallow *Hirundo rustica* (Linnaeus)	C	C	C	—
House Martin *Delichon urbica* (Linnaeus)	S	S	S	—
Sand Martin *Riparia riparia* (Linnaeus)	C	C	C	—
Raven *Corvus corax* (Linnaeus)	S	S	S	S
Carrion/Hooded Crow *Corvus corone* (Linnaeus)	C	C	C	C
Rook *Corvus frugilegus* (Linnaeus)	C	C	C	C
Jackdaw *Corvus monedula* (Linnaeus)	C	C	C	C
Magpie *Pica pica* (Linnaeus)	S	S	S	S
*Great Tit *Parus major* (Linnaeus)	—	—	R	—
*Blue Tit *Parus caeruleus* (Linnaeus)	—	—	R	—
Wren *Troglodytes troglodytes* (Linnaeus)	S	S	S	S
Mistle Thrush *Turdus viscivorus* (Linnaeus)	—	—	S	S
Fieldfare *Turdus pilaris* (Linnaeus)	S	—	S	S

Song Thrush *Turdus philomelos* (Brehm)	S	—	S	S
Redwing *Turdus iliacus* (Linnaeus)	S	—	S	C
Blackbird *Turdus merula* (Linnaeus)	S	S	S	S
Wheatear *Oenanthe oenanthe* (Linnaeus)	C	S	C	—
Stonechat *Saxicola torquata* (Linnaeus)	S	S	S	S
Whinchat *Saxicola rubetra* (Linnaeus)	—	—	R	—
*Redstart *Phoenicurus phoenicura* (Linnaeus)	—	—	R	—
Robin *Erithacus rubecula* (Linnaeus)	S	S	S	S
Sedge Warbler *Acrocephalus schoenobaenus* (Linnaeus)	R	—	S	—
*Whitethroat *Sylvia communis* (Latham)	R	—	R	—
Willow Warbler *Phylloscopus trochilus* (Linnaeus)	R	—	R	—
*Chiffchaff *Phylloscopus collybita* (Vieillot)	R	—	R	—
*Goldcrest *Regulus regulus* (Linnaeus)	R	—	R	—
*Firecrest *Regulus ignicapillis* (Temminck)	—	—	R	—
Dunnock *Prunella modularis* (Linnaeus)	S	S	S	S
Meadow Pipit *Anthus pratensis* (Linnaeus)	C	C	A	C
*Tree Pipit *Anthus trivialis* (Linnaeus)	—	—	R	—
Rock/Water Pipit *Anthus sponoletta* (Linnaeus)	S	S	S	S
Pied/White Wagtail *Motacilla alba* (Linnaeus)	S	—	S	S
Grey Wagtail *Motacilla cinerea* (Tunstall)	S	—	S	S
*Waxwing *Bombycilla garrulus* (Linnaeus)	—	—	—	R
*Great Grey Shrike *Lanius excubitor* (Linnaeus)	—	—	R	—
Starling *Sturnus vulgaris* (Linnaeus)	A	A	A	A
Greenfinch *Carduelis chloris* (Linnaeus)	S	C	C	C
Goldfinch *Carduelis carduelis* (Linnaeus)	C	C	C	C
Siskin *Carduelis spinus* (Linnaeus)	—	—	—	R
Linnet *Acanthis cannabina* (Linnaeus)	C	C	C	C
Twite *Acanthis flavirostris* (Linnaeus)	S	—	S	S
Redpoll *Acanthis flammea* (Linnaeus)	R	—	R	R
Bullfinch *Pyrrhula pyrrhula* (Linnaeus)	—	R	—	—
Chaffinch *Fringilla coelebs* (Linnaeus)	C	C	C	C
Brambling *Fringilla montifringilla* (Linnaeus)	—	—	—	R
Corn Bunting *Emberiza calandra* (Linnaeus)	—	—	R	—
Yellowhammer *Emberiza citrinella* (Linnaeus)	R	—	R	R
Reed Bunting *Emberiza schoenicius* (Linnaeus)	S	S	C	C
*Lapland Bunting *Calcarius lapponicus* (Linnaeus)	—	—	R	—
Snow Bunting *Plectrophenax nivalis* (Linnaeus)	S	—	S	S
House Sparrow *Passer domesticus* (Linnaeus)	C	S	C	C

MARINE FAUNA AND TERRESTRIAL FAUNA
Other than Arachnida and Insecta

Compiled by B. Healy *Department of Zoology University College, Dublin*

Marine fauna

Although the shores of Dublin and the surrounding district have been studied by many generations of marine zoologists and the fauna is on the whole well known, the Bull Island was not often visited by collectors owing, perhaps, to the counter-attractions of other well-known localities such as Dalkey Island and the Malahide Inlet where the fauna is rich and where rare species were more likely to be found. Modern zoologists, however, are generally less interested in discovering new and rare species and more concerned with the ecology and biology of species and in the study of well-defined habitats. In this respect, the Bull Island has presented unique opportunities. Some areas have been studied intensively with the result that the faunas of the mud flat and salt marsh and to a lesser extent the sandy beach are fairly well known. The Bull Wall, on the other hand, has been almost entirely ignored and little is known about the fauna which has colonised the stony shores of the Causeway.

The following list is based mainly on investigations undertaken during the past 20 years and has been compiled from published work, student theses, records in the register of the National Museum and personal observations of local naturalists. Earlier records, dating mainly from the beginning of the century, are sparse, not only due to lack of collecting around the island but also because authors did not always specify localities and were often content to record species as 'from Dublin Bay'. For reasons explained above, the list is biassed in favour of burrowing species, characteristic of mud and sand, while predominantly sessile groups such as Porifera, Cnidaria, Bryozoa and Tunicata and those animals associated with seaweed are poorly represented.

Some specialist groups, i.e. animals which can only be identified by experts in the group, have been studied e.g. Foraminifera, Enchytraeidae, Copepoda, Ostracoda, but others such as Protozoa other than Foraminifera, Turbellaria and Nematoda still await attention.

Terrestrial fauna other than Arachnida and Insecta

Most of the animals included in this category (annelids, woodlice, molluscs and mammals) have relatives which belong to the marine fauna and it has been found convenient therefore to include them in the same list. Some species e.g. *Ligia oceanica* and some of the enchytraeids, span the terrestrial, intertidal interface and could equally well be included in both terrestrial and marine faunas. Terrestrial mammals of course, like birds, range freely over the island including the shoreline.

In the following list, the species appear in alphabetical order within each major group. Names are followed by a letter or letters indicating habitat and a frequency status figure. Sources of records are discussed at the beginning of each section. Personal observations not appearing in theses or published works are referred to by the initials of the observer (see below).

Habitat Code

Marine

A – lagoonal mud flat
B – lagoonal sand flat
C – lagoonal channels
D – *Salicornia* marsh
E – salt marsh channels and pans, lower zone
F – salt marsh channels and pans, upper zone
G – Sutton Spit (sandy mud, vesicular sands)
H – Sutton Creek
I – beach (sand)
J – Sutton Beach (stony)
K – Kilbarrack Beach (stony)
L – Bull Wall

Terrestrial

M – saltmarsh, lower zone
N – saltmarsh, middle zone
O – saltmarsh, upper zone
P – central part of island, old dunes
Q – gardens, rubbish etc.
R – mobile sand dunes
S – young fixed dunes
T – alder marsh
V – Causeway
U – unknown provenance

Frequency status code

+ – only 1 or 2 specimens found
1 – rare
2 – occasional
3 – frequent in suitable habitats but appearances may be erratic
4 – common, at least at certain times of year
5 – common or abundant at all times

Personal observations

BH – B. Healy
BW – A. B. West
JG – J. N. R. Grainger
TOF – T. O'Flaherty
VOB – V. O'Brien
NMI – National Museum of Ireland Register

Protozoa

The only group of Protozoa which has been studied is the Foraminifera for which Harris (1973) records 14 species from sediments. Dead, semi-fossil specimens are ignored but it is not always easy to tell whether specimens are alive or not. Species marked * were probably alive but there is some uncertainty.

Foraminifera

Ammonia beccarii (L.)	B	+
Asterigerinata mamilla (Will.)	I	+
Bulmina gibba/elongata Forn./d'Orb.	I	1
Cyclogyra involvens (Reuss)	A, D	2
Eggerella scabra (Will.)	I	3*
Elphidium articulatum (d'Orb.)	B, C, D	5
Jadammina macrescens (Brady)	E/F	4*

Miliammina fusca (Brady)	E/F	4*
Protelphidium anglicum Murr.	A, B, D	4
Quinqueloculina lata Terq.	A, D	1*
Q. oblongata (Mont.)	E/F	1*
Q. semilunum (L.)	D	1*
Reophax moniliformis Sidd.	E/F	4*
Trochammina inflata (Mont.)	E/F	2

Porifera

All sponges are sessile animals, needing a substratum for an attachment and are therefore probably confined to the Bull Wall. Only one species has been definitely identified (BH) although others are present.

Halichondria panicea (Pall.)	L	3

Cnidaria

Apart from jellyfish (Scyphozoa), which are occasionally washed up on the strand, (one record – BW) Cnidaria are normally sessile and are therefore only to be expected attached to rocks, stones or weed. Several undetermined species of Hydrozoa are present on the Bull Wall.

Hydrozoa		
Dynamena pumila (L.)	L	3
Anthozoa		
Actinia equina (L.)	L	4
Scyphozoa		
Rhizostoma octopus (L.)	I	+

Platyhelminthes

Turbellaria in Dublin Bay have been recorded by Southern (1936) but he does not mention the Bull Island. An orange planarian is very common on the salt marsh (Healy 1975). Several other Turbellaria have been seen and small species are common in most sediments where they constitute an important component of the meiofauna. Tapeworm proglottids (Cestoda) are frequent in salt marsh pans among bird droppings.

Turbellaria		
Uteriporus vulgaris (Berg.)	M, N, O	4

Nemertini

One specimen identified by BW.

Amphiporus lactifloreus (Johnst.)	K	+

Annelida

The marine Polychaeta and Oligochaeta of Dublin Bay are recorded by Southern (1909, 1910) but he does not specifically mention the Bull Island. All marine records in the following list are thus recent. Polychaeta are from Abdel-Moez (1957), Toner (1967), Goodwillie et al. (1970), Healy (1975), BW and NMI (*Pectinaria koreni*). Marine Oligochaeta have only been investigated in the lagoon and salt marsh (Goodwillie et al., 1970, Healy 1975, McGrath 1975). Terrestrial Enchytraeidae have been identified from a few sites (BH) but Lumbricidae, represented on the island by several species, are as yet undetermined. No separation is made of marine and terrestrial Oligochaeta.

Annelida

Polychaeta		
Arenicola marina (L.)	A, B, C, D, E, H, I	5
Capitella capitata (Fabr.)	A, L	2
Cirratulus cirratus (O.F.M.)	L	3
Clymene oerstedii Clap.	J	2
Eteone longa (Fabr.)	B, E	3
Eulalia sanguinea (Oerst.)	H	+
Fabricia sabella (Ehrenb.)	A, E	3
Glycera convolua Kefers.	H	3
Harmothoe impar (Johns.)	H	+
Lanice conchilega (Pallas)	B, H, I	4
Manyunkia aestuarina (Bourne)	E	3
Nephtlys caeca Fabr.	I	4
N. cirrosa Ehlers	I	+
N. hombergi Aud. and M-E.	H, I, K	3
N. incisa Malmg.	B	3
Nereis diversicolor O.F.M.	A, B, C, E, F	4
N. pelagica L.	B	1
N. virens M. Sars	I	+
Nerine cirratulus (Delle Chiaje)	I	1
Paraonis fulgens (Levins.)	H	1
Pectinaria koreni Malmg.	I	+
Phyllodoce laminosa Sav.	B	2
Pygospio elegans Clap.	B, E, I	3
Scolecolepis fuliginosa (Clap.)	A	1
Scoloplos armiger (O.F.M.)	A, B, I	2
Spio filicornis (O.F.M.)	B, H, I, K	4
Spirorbis borealis Daud.	J	4
S. pagenstecheri Quatref.	J	4
S. rupestris Gee and K-J	J	2

Oligochaeta

Achaeta littoralis Lass.	S	3
Bryodrilus parvus Nurm.	S	2
Buchholzia appendiculata (Buch.)	S	2
B. fallax Mich.	T	3
Clitellio arenarius (O.F.M.)	A, B, E	3
Enchytraeus albidus Henle	N, O	4
E. buchholzi Vejd.	O, P, R, S, T	3
E. capitatus v. Bul.	R	3
E. minutus N. and C.	S	2
Fridericia bulbosa (Rosa)	S, T	3
F. callosa (Eisen)	O, S, T	4
F. gracilis v. Bul.	S	3
F. perrieri (Vejd.)	T	2
F. ratzeli (Eisen)	S, T	2
Henlea nasuta (Eisen)	S, T	3
H. perpusilla Friend	T	2
H. ventriculosa (Udek.)	S, T	3
Limnodrilus hoffmeisteri Clap.	E	4
Lumbricillus kaloensis N. and C.	A, E, M	4
Marionina achaeta Lass.	O, R	3
M. appendiculata N. and C.	A, E, M	3
M. argentea (Mich.)	M, O, R, S, T	3
M. communis N. and C.	S	3
M. libra N. and C.	S	2
M. spicula (Leuck.)	N, O, R S	5
M. subterranea (Knölln.)	M, O, R	2
Monopylephorus rubroniveus Levins.	F	2
Paranais littoralis (Müll.)	A	3
Peloscolex benedeni (Udek.)	A, B, C, D, E	5
Stylodrilus heringianus Clap.	E	
Tubifex costatus Clap.	A, D, E, F	5
T. pseudogaster (Dahl)	A	1

Crustacea

Records of Copepoda have been summarised by O'Riordan (1971) who himself collected widely on Bull Island. Ostracoda (live specimens only) are from Harris (1973). The marine Isopoda were mainly collected by Brandt (1971); terrestrial species have been the subject of a recent unpublished survey by D. Dooge. Amphipoda and Decapoda are recorded by Goodwillie et al. (1970), Healy (1975), BW and JG.

The most interesting species of crustacean is the woodlouse *Eluma purpurascens*, first discovered by Pack-Beresford (1908) and Bagnall (1908) for which the only known localities in the British Isles apart from the Bull Island are Howth and Portmarnock.

Crustacea

Ostracoda

Cytherois fischeri (Sars)	A, D, E	2
Leptocythere lacertosa (Hirschm.)	A, B, D, E	4
L. pellucida (Baird)	A, B, D, E	3
Pontocythere elongata (Brady)	H, I	1

Copepoda

Amphiascoides nana (Sars)	A	+
Asellopsis intermedia (T. Scott)	B, I, K	5
Canuella furcigera Sars	I	1
C. perplexa T. A. Scott	A, I	3
Diarthrodes nobilis Baird	I	+
Enhydrosoma buchholtzi (Boeck)	E	1
E. propinquum (Brady)	A	1
Eurytemora hirudinoides (Nordq.)	I	1
Halectinosoma herdmani (T. and A. Scott)	I	4
Harpacticus flexus Br. and Rob.	I	4
H. littoralis T. and A. Scott	E	2
Heterolaophonte littoralis (T. and A. Scott)	E	1
H. minuta (Boeck)	E, K	1
H. stroemi (Baird)	B	1
Itunella muelleri (Gag.)	M	3
Laophonte setosa Boeck	E	+
Mesochra aestuarii Mon.	A	+
M. heldti Mon.	B, E	1
M. lilljeborgi Boeck	A, E	3
Nannopus palustris Brady	E	+
Paranychocamptus curticaudatus Boeck	A, B, E, K	2
Parathalestris intermedia Gurn.	A, E	3
Platychelipes littoralis Brady	A, B, E, K	5
Pseudobradya attenuata T. and A. Scott	E	+
Robertsonia celtica (Mon.)	B, K	1
Stenhelia palustris Brady	A, E, I	4
Tachidius discipes Giesb.	A, B, E	3
Thomsonula hyenae I.C. Thom.	I	5

Cirripedia

Balanus balanoides (L.)	D, G, K, L	4

Isopoda—marine

Paragnathia formica (Hesse)	M	3
Eurydice pulchra Leach	E, I	5
Idotea granulosa Rathke	L	1
I. chelipes (Pallas)	B	+
Jaera albifrons Leach	K	3–5

Isopoda—terrestrial		
Androniscus dentiger Verh.	V	1
Armadillidium vulgare (Latr.)	O, P	5
Eluma purpurascens (B.-L.)	O, P	5
Ligia oceanica (L.)	L, V	3
Metaponorthus cingendus (Kin.)	V	+
Oniscus asellus (L.)	O, P	4
Philoniscus muscorum (Scop.)	O, P	4
Porcellio scaber (Latr.)	O, P	4
Trichoniscus pusillus (Brandt)	T, V	2
T. pygmaeus (Sars)	T	1

Amphipoda

Bathyporeia pelagica Bate	I	2–3
B. pilosa Lindstr.	I	2–3
Corophium volutator (Pallas)	A, B, C, D, E, F	5
Gammarus zaddachi Sext.	E	3
Haustorius arenarius (Slab.)	I	2
Orchestia gammarella (Pallas)	E, F, M, N, O	4
Urothoe marina (Bate)	I	2–3

Decapoda

Carcinus maenas (L.)	A, B, C, E, F, J, K, L, M	5
Corystes cassivelaunus Penn.	I	+
Crangon vulgaris Fabr.	C, E, F, I	4
Hyas araneus (L.)	U	+
H. coarctatus Leach	U	+
Palaemonetes varians (Leach)	E	2
Porcellana longicornis (L.)	U	+
Pinnotheres pisum Penn.	J	2

Diplopoda

Cylindroiulus latestriatus (Curt.)	O	4

Mollusca

The sandy beach has been a favourite haunt of shell collectors for more than a century and many exciting finds have been made. Empty shells are not recorded here however because their provenance is unknown and they do not strictly belong to the island. Recently dead, stranded animals are included however. Apart from isolated records of dead squid (NMI), all the records are recent because neither Nichols (1910) nor Colgan (1914, 1930) mention the Bull Island in their publications. The list for Lamellibranchia is probably more or less complete (Goodwillie et al. 1970, Harris 1973, Healy 1975, Goodman 1976, BW). Marine Gastropoda are more spasmodic in their appearance and many occur on rocks or seaweed, so these are less well known. The land shells (*Pulmonata*) have been investigated by D. Dooge who has made some interesting observations on the behavioural responses to drought of the dominant species.

Mollusca

Gastropoda—Marine		
Acanthodoris pilosa (O.F.M.)	L	+
Acmaea testudinalis Müll.	E	+
Akera bullata Müll.	E	+
Alderia modesta (Loren)	A, E	2
Hydrobia ulvae (Penn.)	A, B, C, D, E, F, M	5
Limapontia depressa Ald. and Han.	A, E	3
Littorina littoralis (L.)	E, K, L	2
L. littorea (L.)	A, E, K	3
L. saxatilis (Olivi)	A, B, C, D, E, F, H, J, K, L	5
Nucella lapillus (L.)	L	3
Patella vulgata (L.)	L	4

Gastropoda—terrestrial

Arion ater agg.	P, Q	1
A. subfuscus (Drap.)	P, Q	1
Cochlicella acuta (Müll.)	P	5
Cochlicopa lubrica (Müll.)	P, T	2
C. lubricella (Porro)	R, S, P,	1
	T	3
Discus rotundatus (Müll.)	P	+
Helicella caperata (Mont.)	R, S	2
H. virgata (da Costa)	R, S, P	5
Helix aspersa (Müll.)	R, S, P, Q	5
Hygromia striolata (C. Pfeiff.)	Q	1
Milax budapestensis (Hazay)	Q	1
M. gagates (Drap.)	Q	1
M. sowerbi (Ferus.)	Q	1
Oxychilus alliarius (Mill.)	P	3
Phytia myosotis Drap.	M, N, O	4
Pupilla muscorum (L.)	S	4
Retinella nitidula (Drap.)	P	2
R. radiata (Alder)	P	2
Vallonia excentrica Sterki	S, P	3
V. pulchella (Müll.)	T	1
Vitrina pellucida (Müll.)	S, P, Q	3

Lamellibranchia		
Acanthocardia echinata (L.)	I	+
Arctica islandica (L.)	U	+
Cardium edule (L.)	A, B, C, E, I	4
Macoma balthica (L.)	A, B, I	4
Mya arenaria (L.)	A	2
Mytilus edulis (L.)	G, L	5
Scrobicularia plana (da Costa)	A, E	3
Tellina tenuis (da Costa)	I	4

Pisces

There is no lack of published information on the fish of Dublin Bay but few authors indicate localities. Shore species are recorded by Bracken and Kennedy (1967), Kunz (1969), Goodwillie et al. (1970), Healy (1975) and BW. Apart from the two species of sea horse, which were found stranded on the beach, (Thompson 1856, More 1878), and a few species listed by O'Riordan (1966), records of offshore fish are mainly due to the observations of local anglers (VOB and TOF).

Pomatoschistus microps is seasonally abundant in the salt marsh and lagoon where juvenile mullet and elvers may also be common at certain times. For most species however it is impossible to give any indication of abundance.

Pisces – Fish

Ammodytes tobianus (L.)—lesser sand eel	G
Ammodytes spp.—sand eels	C, I
Anguilla anguilla (L.)—eel	E, F, L
Belone belone (L.)—garfish	L
Blennius pholis (L.)—shanny	G
Ciliata mustela (L.)—5-bearded rockling	G
Conger conger (L.)—conger eel	L
Crenimugil labrosus (Risso)—thick-lipped mullet	C, E, F, L, I
C. melops (L.)—gilt-head	G
Dicentrarchus labrax (L.)—bass	C, L, I
Gadus morhua (L.)—cod	C, L, I
Hippocampus antiquorum Leach—sea horse	U
H. brevirostris Cuv.—sea horse	U
Labrus bergylta Asc.—ballan wrasse	L
Lepidorhombus wiff-iagonis (Wal.)—megrum	C, I
Limanda limanda (L.)—dab	C, I
Merlangus merlangus (L.)—whiting	C, L, I
Pholis gunnellus (L.)—butterfish	G
Phrynorhombus regius (Bonn.)—Eckstrom's topknot	C, I
Platichthys flesus (L.)—flounder	C, L, I
Pleuronectes platessa (L.)—plaice	C, L, I
Pollachius pollachius (L.)—pollack	C, L, I
P. virens (L.)—coalfish	L
Pomatoschistus microps Kroy.—small goby	C, E, F
Salmo trutta (L.)—sea trout	L
Scomber scombrus (L.)—mackerel	L
Scyliorhinus canicula (L.)—lesser-spotted dogfish	L
Spinachia spinachia (L.)—15-spined stickleback	G
Syngnathus acus (L.)—greater pipefish	L
S. rostellatus Nilss.—Nilsson's pipefish	G
Taurulus bubalis (Euph.)—long-spined sea scorpion	G
Trachinus vipera Cuv. and Val.—lesser weever	C, L, I
Trisopterus luscus (L.)—pouting	C, L, I
Zeugopterus punctatus (Bloch)—common topknot	C, I

Mammals

The following list is remarkable chiefly for its brevity which may be attributable to a lack of reliable (or communicative) observers rather than to the absence of species. The fact that there are no recent records of rabbits – apparently plentiful in the past, (O'Mahoney 1931) – is particularly astonishing, especially in view of the undoubted presence of burrows in the dunes. While the absence of shelter may deter some species from making permanent homes on the island, the Causeway, with its stony shoreline and established vegetation, now constitutes a connection with the mainland over which foxes, shrews and even stoats and badgers may be expected to venture, at least occasionally. It may be only a matter of time before sightings of these species are reported.

The three records of marine species are taken from the register of the National Museum of Ireland and refer to strandings.

Mammalia

Apodemus sylvaticus (L.)—wood (field) mouse
Halichoerus grypus (Fab.)—grey seal
Lagenorhynchus albirostris (Gray)—white beaked dolphin
Lepus timidus hibernicus Bell—Irish hare
Phocaena phocaena (L.)—common porpoise
Mus musculus (L.)—house mouse
Rattus norvegicus (Erxl.)—brown rat

ARACHNIDA and INSECTA

Compiled by M. C. D. Speight *Department of Zoology Trinity College*
& B. Healy *Department of Zoology University College Dublin*

Some 420 species are recorded in the following list. Perhaps this sounds a lot of species, but in reality it can only represent a fraction of the Bull Island invertebrate fauna. Some large insect groups like the Hymenoptera (ants, bees, wasps, ichneumons, sawflies etc.) are very poorly known from the North Bull. For instance, there are as yet no wasp (*Vespula* spp.) records at all. On the other hand, it is equally true that very few lists of invertebrates from costal dune localities in the British Isles have been published – the Bull Island list is arguably the most comprehensive published so far.

Many of the records in this list are drawn from other publications, some dating from very early in this century. We have not been able to check the validity of these records. It must also be pointed out that changes which have occurred in the character of Bull Island during this century will have affected the ability of various species to survive there – some of those recorded in this list may no longer occur on the North Bull.

In addition to listing the names of the invertebrates known from Bull Island we have attempted to give details of date of capture, habitat, collector's name, determiner's name and, where relevant, the publication from which each record is derived.

Unfortunately, early collectors often failed to give details of date and habitat. Since it proved necessary to use abbreviations to refer to habitats and people's names in the list, a key to these abbreviations precedes the list proper. It should be noted that where a species was both collected and determined by the same person, only one set of initials follows the habitat symbol, e.g. (BH), but where an animal was collected by one individual and determined by someone else two sets of initials are given, e.g. (BH/MS), the second set being those of the determiner.

The papers used in compiling the list are to be found in the bibliography. We have not used publications from the last century, but some 19th century records have turned up in later publications. These latter records are included in the list with the year of capture given in full, e.g. 1898. In records from the 20th century, the year of capture is occasionally abbreviated, e.g. 75. Where a species has been recorded more than once, only the most recent record is listed.

Names of organisms given in the list are as given in the most up-to-date check lists at our disposal and are thus not necessarily as they appear in earlier publications. Families are listed in an order which supposedly reflects their phylogenetic relationships as understood currently, but within each family the genera and species are listed alphabetically, for ease of reference.

Key to abbreviations used

Column one—**capture date**
? = capture date unknown

Column two—**habitat**
AD = *Ammophila* dunes
BW = bull wall
C = causeway
DG = dune grassland
DS = dune slack (alder marsh)
G = generally distributed
SM = salt marsh
U = provenance unknown

Column three—**collectors and determiners**
AS = A. W. Stelfox
B = E. W. Bullock
BH = B. Healy
C = P. C. Chandler
Cr = R. W. Crosskey
D = J. Deeming
DD = D. Doogue
DH = D. Hollis
DK = D. E. Kimmins
DS = J. F. Dawson
E = G. O. Evans
F = O. S. Fletcher
FS = F. R. Speyer
G = M. W. R. de V. Graham
H = J. N. Halbert
Ha = A. H. Haliday
Hn = A. R. Hogan
J = J. C. Jerrard
JB = J. D. Bradley
K = (see M. Speight) Kemp
L = P. N. Lawrence
KS = K. G. V. Smith
M = W. R. McNab
MGM = M. G. Morris

ML	=M. Luxton
MS	=M. C. D. Speight
N	=R. Nash
O	=H. Oldroyd
O'M	=E. O'Mahony
P	=C. F. Prevost
P-B	=D. R. Pack-Beresford
PH	=P. Hillis
S	=(see M. Speight) Smit

Column four—**date of publication as cited in bibliography**
(in pr.)=in press
(n. r.) =new record published for the first time in this list.

ARACHNIDA

Acari

Oribatei

Ameronothrus bilineatus (Mich.)	1960/3	**SM**	BH/E	1975
A. lineatus brevipes Will.	1960/3	**SM**	BH/ML	1975
Hermannia reticulata Thor.	1960/3	**SM**	BH/ML	1975
H. subglabra Berl.	1960/3	**SM**	BH/E	1975
Hygroribates bilineatus Bost.	1960/3	**SM**	BH/ML	1975
H. nigrofemoratus (L. Koch)	1960/3	**SM**	BH/ML	1975
H. spoofi (Oud.)	1960/3	**SM**	BH/ML	1975
Liebstadia similis (Mich.)	1960/3	**SM**	BH/ML	1975
Oppia clavipectinata (Mich.)	1960/3	**SM**	BH/ML	1975
Platynothrus peltifer (C. L. Koch)	1960/3	**SM**	BH/ML	1975
Punctoribates quadrivertex Halb.	1960/3	**SM**	BH/ML	1975
Scheloribates laevigatus (C. L. Koch)	1960/3	**SM**	BH/ML	1975
Trichoribates incisellus (Kram.)	1960/3	**SM**	BH/E	1975

Mesostigmata

Blattosocius dendriticus (Berl.)	1960/3	**SM**	BH/E	1975
Cheiroseius necorniger (Oud.)	1960/3	**SM**	BH/ML	1975
Digamasellus halophilus (Will)	1960/3	**SM**	BH/E	1975
Leioseius salinus Halb.	1960/3	**SM**	BH/E	1975
Marocheles scutatus (Berl.)	1960/3	**SM**	BH/ML	1975
Pergamasus longicornis Berl.	1960/3	**SM**	BH/E	1975
Phaulodinychus orchestiidarum Barr.	1960/3	**SM**	BH/E	1975
P. minor (Halb.)	1960/3	**SM**	BH/ML	1975
Rhodacarus pallidus Hull	1960/3	**SM**	BH/ML	1975

Prostigmata

Bdella longicornis (L.)	1960/3	**SM**	BH/E	1975
Cheylostigmaeus salinus Evans	1960/3		BH/ML	1975
C. scutatus (Halb.)	1960/3	**SM**	BH/E	1975
Eupodes halophilus Halb.	1960/3	**SM**	BH/E	1975

Araneae

Dictynidae

Argenna subnigra (O.P.-C.)	1960/3	**SM**	BH/J	1975

Dysderidae

Clubiona stagnatilis Kulcz.	1960/3	**SM**	BH/J	1975
Segestria senoculata (L.)	1960/3	**SM**	BH/J	1975

Thomisidae

Thanatus striatus (C. L. Koch)	1960/3	**SM**	BH/J	1975
Tibellus oblongus Walck.	1960/3	**SM**	BH/J	1975
Xysticus cristatus (Clerck)	1960/3	**SM**	BH/J	1975

Lycosidae

Lycosa pullata (Clerck)	1960/3	**SM**	BH/J	1975
L. purbeckensis (O.P.-C)	1960/3	**SM**	BH/J	1975
Pirata piratica Clerck	1960/3	**SM**	BH/J	1975
Trochosa terricola Thorell	1960/3	**SM**	BH/J	1975

Theridiidae

Robertus lividus (Blackw.)	1960/3	**SM**	BH/J	1975

Tetragnathidae

Pachygnatha clercki Sund.	1960/3	**SM**	BH/J	1975
P. degeeri Sund.	1960/3	**SM**	BH/J	1975
Tetragnatha extensa (L.)	1960/3	**SM**	BH/J	1975

Araneidae

Araneus cornutus Clerck	1960/3	**SM**	BH/J	1975
A. diadematus Clerck	1960/3	**SM**	BH/J	1975
A. quadratus Clerck	1960/3	**SM**	BH/J	1975

Linyphiidae

Bathyphantes gracilis (Blackw.)	1960/3	**SM**	BH/J	1975
Erigone longipalpis (Sund.)	1960/3	**SM**	BH/J	1975
Hypomma bituberculatum (Wid.)	1960/3	**SM**	BH/J	1975
Lepthyphantes tenuis (Blackw.)	1960/3	**SM**	BH/J	1975
Meionata rurestris Koch	Sept 1898	U	P–B	1909
Mengea scopigera (Grube)	1960/3	**SM**	BH/J	1975
Oedothorax fuscus (Blackw.)	1960/3	**SM**	BH/J	1975
Savignia frontata (Blackw.)	1960/3	**SM**	BH/J	1975
Tapinocyba praecox (O.P.-C.)	Feb 1927	U	P–B	1929
Typhlocrestus digitatus (O.P.-C.)	1910	U	P–B	1929

Phalangida

Phalangidae

Phalangidium opilio (L.)	1960/3	**SM**	BH	1975

INSECTA

Collembola

Hypogastruridae

Anurida maritima Guerin	1960/3	**SM**	BH/L	1975
Friesia mirabilis (Tullb.)	1960/3	**SM**	BH/L	1975
Xenylla (Fab.)	1960/3	**SM**	BH/L	1975

Onchiuridae

Onychiurus debilis (Moniez)	1960/3	**SM**	BH/L	1975
Tullbergia krausbaueri (Börner)	1960/3	**SM**	BH/L	1975

Isotomidae

Folsomia sexoculata (Tullb.)	1960/3	**SM**	BH/L	1975
Isotoma maritima (Tullb.)	1960/3	**SM**	BH/L	1975
I. viridis Bourl.	1960/3	**SM**	BH/L	1975
Isotomiella minor (Schaff.)	1960/3	**SM**	BH/L	1975

Dermaptera

Forficulidae

Forficula auricularia L.	1960/3	**SM**	BH	1975

Psocoptera

Polypsocidae

Caecilius burmeisteri Brauer	1960/3	**SM**	BH/DK	1975

Pseudocaecilidae

Ectopsocus briggsi (MacL.)	1960/3	**SM**	BH/DK	1975

Thysanoptera

Thripidae

Aptinothrips nitidulus Hal.	1960/3	**SM**	BH/FS	1975
A. stylifer Tryb.	1960/3	**SM**	BH/FS	1975
Haplothrips juncorum Bag.	1960/3	**SM**	BH/FS	1975
H. statices (Hal.)	1960/3	**SM**	BH/FS	1975
Chirothrips manicatus Hal.	1960/3	**SM**	BH/FS	1975
Taeniothrips atratus (Hal.)	1960/3	**SM**	BH/FS	1975
Thrips tabaci Lind.	1960/3	**SM**	BH/FS	1975

Hemiptera Heteroptera

Lygaeidae

Nysius thymi (Wolff)	12. VII. 75	**DS**	BH/MS	(n.r.)
Scolopostethus affinis Schill.	?	**U**	H	1935
Stygnocoris fuliginosus Geoff.	VII. 1898	**U**	H	1935
S. pedestris Fal.	VIII. 1893	**U**	H	1935
S. rusticus (Fal.)	1960/3	**SM**	BH	1975

Nabidae

Nabis ferus L.	IX. 1898	**U**	BH	1935
N. flavomarginatus Sch.	1960/3	**SM**	BH	(n.r.)

Cimicidae

Anthocoris nemorum (L.)	1963/3	**SM**	BH	1975

Saldidae

Chiloxanthus pilosa (Fal.)	7. VIII. 1898	**U**	H	1935
Salda littoralis L.	1960/3	**SM**	BH	1975
Saldula orthochila Fieb.	7. VIII. 1898	**U**	H	1935
S. palustris (D. & S.)	1960/3	**SM**	BH	1975
S. saltatoria L.	?	**U**	H	1935

Miridae

Calocoris norvegicus (Gmel.)	1960/3	**SM**	BH	1975
Fieberocapsus flaveolus (Reut.)	1960/3	**SM**	BH	1975
Leptoterna dolabrata (L.)	12. VII. 75	**DS**	BH/MS	(n.r.)
L. ferrugata Fal.	1960/3	**SM**	BH	1975
Macrotylus paykulli Fal.	VII. 1898	**U**	H	1935
Plagiognathus arbustorum Fab.	VI. 1893	**U**	H	1935
P. chrysanthemi Wol.	VIII. 28	**U**	H	1935
Teratocoris saundersi D. & S.	VII. 1898	**U**	H	1935
Trigonotylus ruficornis Geoff.	VII. 1898	**U**	H	1935

Hemiptera Homoptera

Cercopidae

Neophilaenus exclamationis (Thunb.)	12. VII. 75	**DS**	BH/MS	
N. lineatus (L.)	12. VII. 75	**DS**	BH/MS	
Philaenus spumarius (L.)	12. VII. 75	**DS**	BH/MS	

Cicadellidae

Aphrodes bicinctus (Sch.)	1960/3	**SM**	BH	1975
A. histrioncus Fab.	?	**U**	H	1935
Arthaldeus pascuellus (Fal.)	1960/3	**SM**	BH	1975
Euscelis obsoletus (Kirsch.)	1960/3	**SM**	BH	1975
Macrosteles horvarthi (Wag.)	1960/3	**SM**	BH	1975
Psammotettix sabulicola Curt.	?	**U**	H	1935

Delphacidae

Javesella dubia (Kirsch.)	?	**SM**	BH	1975
Microdelphax aubei Perris	?	**AD**	H	1935

Pysllidae

Psylla foersteri Flor.	12. VII. 75	**DS**	BH/DH	(n. r.)

Callaphididae

Juncobium leegei (Born.)	1960/3	**SM**	BH/P	1975

Aphididae

Aphis tripolii (Laing)	1960/3	**SM**	BH/P	1975
Brachycaudus helichrysi (Kalt.)	1960/3	**SM**	BH/P	1975
Macrosiphoniella asteris (Walk.)	1960/3	**SM**	BH/P	1975
Macrosiphum avenae (Fab.)	1960/3	**SM**	BH/P	1975
M. euphorbiae (Thomas)	1960/3	**SM**	BH/P	1975
M. rosae (L.)	1960/3	**SM**	BH/P	1975
Metapolophium dirhodum (Walk.)	1960/3	**SM**	BH/P	1975
Myzus cerasi (Fab.)	1960/3	**SM**	BH/P	1975
Rhopalosiphum padi (Schr.)	1960/3	**SM**	BH/P	1975

Pemphigidae

Aploneura lentisci Passerini	1960/3	**SM**	BH/P	1975

Trichoptera

Limnephilidae

Limnephilus affinis Curt.	1960/3	**SM**	BH	1975

Lepidoptera

Hypsidae

Hypocrita jacobaeae (L.)	1975	**AD**	BH	(n.r.)

Caradrinidae

Amphipoea lucens (Frey.)	1960/3	**SM**	BH/F	1975
Lacanobia suasa (D & S)	1960/3	**SM**	BH	1975
Leucania impura (Hueb.)	1960/3	**SM**	BH	1975

Lycaenidae

Polyommatus icarus (v. Rott.)	1975	**DG**	BH	(n.r.)

Pieridae

Pieris brassicae (L.)	1975	**DG**	BH	1975

Satyridae

Coenonympha pamphilus (L.)	?	**DG**	PH	(n.r.)

Nymphalidae

Aglais urticae (L.)	1975	**DG**	BH	(n.r.)

Zygaenidae

Zygaena filipendulae (L.)	1975	**DG**	BH	1975

Pyralidae

Chrysoteuchia hortuella (Hub.)	1960/3	**SM**	BH/JB	1975
Crambus perlella (Scop.)	1960/3	**SM**	BH/JB	1975

Cochylidae

Phalonidia affinitana (Doug.)	1960/3	**SM**	BH/JB	1975

Gelechidae

Scrobipalpa samadensis (Stain.)	1960/3	**SM**	BH/JB	1975

Coleophoridae

Coleophora alticolella (Zell.)	1960/3	**SM**	BH/JB	1975
C. maritimella Newm.	1960/3	**SM**	BH/JB	1975

Neuroptera

Chrysopidae

Chrysopa ventralis Curt.	12. VII. 75	**DS**	BH/MS	(n.r.)

Anoplura

Haemotopinidae

Polyplax serrata (Burm.)	?	**U**	O'M	1947
P. spinulosa (Burm.)	?	**U**	O'M	1947
Haemodipsus lyriocephalus (Burm.)	?	**U**	O'M	1947

Coleoptera

Carabidae

Aepus marinus Strom.	?	**U**	Hn	1902
Amara aenea (Deg.)	1960/3	**SM**	BH	1975
A. praetermissa C. R. Sahlb.	?	**U**	DS	1902
A tibialis Payk.	2. V. 75	**AD**	MS	(n.r.)
Calathus fuscipes Goez.	1960/3	**SM**	BH	1975
C. melanocephalus (L.)	1960/3	**SM**	BH	1975
C. mollis Marsh	IV. 25	**U**	B	(n.r.)
Demetrias atricapillus L.	14. II. 71	**AD**	MS	(n.r.)
Dicheirotrichus gustavi Crotch	1960/3	**SM**	BH	1975
Dromius linearis (Oliv.)	1960/3	**SM**	BH	1975
D. melanocephalus L.	14. II. 75	**AD**	MS	(n.r.)
Dyschirius globosus (Herbst)	1960/3	**SM**	BH	1975
Pristonychus complanatus Dj.	1902	**U**	K	1929
P. terricola Herbst.	1960/3	**U**	H	1902
Pterostichus strenuus (Panz.)	1960/3	**SM**	BH	1975
P. versicolor Sturm	1960/3	**SM**	BH	1975
Trechus obtusus Erich.	4. IV. 71	**DS**	MS	(n.r.)
T. rubens Fab.	?	**U**	Ha	1902

Hydrophilidae

Helophorus minutus Fab.	1960/3	**SM**	BH	1975
Ochthebius auriculatus Rey	1900	**U**	H	1902
O. dilatatus Steph.	1963	**SM**	BH	1975
O. punctatus Steph.	1960/3	**SM**	BH	1975
O. marinus (Payk.)	1960/3	**SM**	BH	1975

Silphidae

Necrophorus vespillo (L.)	1960/3	**SM**	BH	1975
Silpha tristis Ill.	1960/3	**SM**	BH	1975
Thanatus rugosus (L.)	1960/3	**SM**	BH	1975

Staphilinidae

Acidota crenata (Fab.)	1960/3	**SM**	BH/OWS	1975
Atheta atramentaria (Gyll.)	1960/3	**SM**	BH/OWS	1975
A. fungi (Grav.)	1960/3	**SM**	BH/OWS	1975
A. flavipes (Thom.)	1960/3	**SM**	BH/OWS	1975
A. hypnorum (Kies.)	1960/3	**SM**	BH/OWS	1975
A. mortuorum Thom.	1900	**U**	H	1902
Bledius arenoides Tott.	1900	**U**	H	1902
B. opacus (Block)	1900	**U**	H	1902
B. spectabilis Kr.	1960/3	**SM**	BH	1975
Diglotta mersa (Hal.)	?	**U**	Ha	1902

Gyrohypnus laeviusculus (Steph.)	?	**U**	H	1902
Heterothops binotatus (Grav.)	1960/3	**SM**	BH/OWS	1975
Hypocyptus longicornis (Payk.)	?	**U**	H	1902
Lamprinodes saginatus (Grav.)	1960/3	**SM**	BH/OWS	1975
Megarthrus depressus (Payk.)	1960/3	**SM**	BH/OWS	1975
Mycetoporus splendidus (Grav.)	1960/3	**SM**	BH/OWS	1975
Omalium laeviusculum Gyll.	1960/3	**SM**	BH/OWS	1975
O. rugulipenne Rye	?	**U**	K	1929
Oxypoda waterhousei Rye	1900	**U**	H	1902
Oxytelus perrisi Fauv.	1900	**U**	H	1902
Philonthus albipes (Grav.)	1900	**U**	H	1902
P. fuscipennis (Mann.)	1960/3	**SM**	BH/OWS	1975
P. marginatus (Fab.)	1960/3	**SM**	BH/OWS	1975
P. varians (Payk).	1960/3	**SM**	BH/OWS	1975
Staphilinus ater Grav.	1960/3	**SM**	BH/OWS	1975
Tachinus rufipes (DeG.)	1960/3	**SM**	BH/OWS	1975
Tachyporus chrysomelinus (L.)	1960/3	**SM**	BH/OWS	1975
T. hypnorum (Fab.)	1960/3	**SM**	BH/OWS	1975
T. nitidulus (Fab.)	1960/3	**SM**	BH/OWS	1975
T. pusillus Grav.	1960/3	**SM**	BH/OWS	1975

Histeridae

Hister purpurascens Herbst.	1900	**U**	H	1902
Pachylopus maritimus (Steph.)	1900	**U**	H	1902

Cantharidae

Cantharis thoracica (Oliv.)	1960/3	**SM**	BH	1975
C. darwiniana (Sharp)	1960/3	**SM**	BH	1975
C. fulvicollis (Fab.)	1960/3	**SM**	BH	1975

Dasytidae

Psilothrix cyaneus (Oliv.)	1960/3	**SM**	BH	1975

Elateridae

Agriotes lineatus (L.)	1960/3	**SM**	BH	1975

Heteroceridae

Heterocerus flexuosus Steph.	1960/3	**SM**	BH	1975
Omosita discoidea (Fab.)	?	**SM**	O'M	1924

Nitidulidae

Meligethes erythropus (Gyll.)	1900	**U**	H	1902

Cryptophagidae

Antherophagus nigricornis (Fab.)	?	**U**	O'M	1929
Cryptophagus cellaris (Scop.)	1900	**U**	H	1902

Lathridiidae

Corticaria crenulata (Gyll.)	1902	**U**	H	1902

Coccinellidae

Coccinella 7–punctata L.	1960/3	**SM**	BH	1975
C. 11–punctata L.	14. II. 71	**SM**	MS	(n.r.)

Scarabeidae

Aphodius aestivalis Steph.	1900	**U**	H	1902

Chrysomelidae Chaetocnema

Chaetocnema concinna (Marsh.)	1960/3	**SM**	BH	1975
Longitarsus ochroleucus (Marsh.)	?	**U**	H	1902

Curculionidae

Apion dichroum Bedel	1960/3	**SM**	BH	1975
A. rubens Steph.	?	**U**	H	1937
Notaris bimaculatus (Fab.)	4. IV. 71	**SM**	MS/ MGM	(n.r.)
Otiorrhyncus atroapterus (DeG.)	2. V. 76	**AD**	MS	(n.r.)
O. ligneus (Oliv.)	1960/3	**SM**	BH	1975
Philopedon plagiatus (Schal.)	2. V. 76	**AD**	MS	(n.r.)
Polydrusus chrysomela (Oliv.)	1960/3	**SM**	BH	1975

Hymenoptera Symphyta

Tenthredinidae

Heterarthrus vagans (Fal.)	26. VI. 76	**DS**	MS	(n.r.)
Tenthredopsis nassata (L.)	26. VI. 76	**DS**	MS	(n.r.)

Hymenoptera Apocrita

Apidae

Apis mellifera L.	1960/3	**SM**	BH	1975
Bombus jonellus (Kirby)	2. V. 76	**AD**	MS	(n.r.)
B. muscorum (L.)	12. VII. 76	**DG**	BH/MS	(n.r.)
B. ruderarius (Mull.)	26. VI. 76	**DS**	MS	(n.r.)

Formicidae

Myrmica rubra (L.)	1960/3	**SM**	BH	1975
M. ruginodis Nyl.	12. VII. 76	**AD**	BH/MS	(n.r.)

Pompilidae

Psammochares plumerus Fab.	7. VIII. 1898	**AD**	H	1927

Sphecidae

Crossocerus wesmaeli v. d. Lind	7. VIII. 1898	**AD**	H	1927
Psen unicolor v. d. Lind	7. VIII. 1898	**U**	H	1927

Vespidae

Ancistrocerus pictus Curt.	July	**U**	O'M	1927

Hymenoptera Parasitica

Braconidae

Apantales triangulator (Wesm.)	1960/3	**SM**	BH/AS	1975
Aphaerata minuta (Hal.)	1960/3	**SM**	BH/AS	1975
Aspilota concolor (Nees)	1960/3	**SM**	BH/AS	1975
Bracon anthracinus Nees	1960/3	**SM**	BH/AS	1975
Dacnusa anasella Stelf.	1960/3	**SM**	BH/AS	1975
D. senilis (Nees)	1960/3	**SM**	BH/AS	1975
Microgaster globata Nees	1960/3	**SM**	BH/AS	1975
Microplitis xanthopus (Ruthe)	1960/3	**SM**	BH/AS	1975
Rhizarcha areolaris (Nees)	1960/3	**SM**	BH/AS	1975

Aphidiidae

Aphidius matricariae Hal.	1960/3	**SM**	BH/AS	1975

Ichneumonidae

Angitia cerophaga (Grav.)	1960/3	**SM**	BH/AS	1975
A. exareolata (Grav.)	1960/3	**SM**	BH/AS	1975
Atractodes tenebriodes (Grav.)	1960/3	**SM**	BH/AS	1975
Diplazon laetatorius (Fab.)	1960/3	**SM**	BH/AS	1975
D. signatus (Grav.)	1960/3	**SM**	BH/AS	1975
Ephialtes brevicornis (Grav.)	1960/3	**SM**	BH/AS	1975
Gelis anthracina (Foer.)	1960/3	**SM**	BH/AS	1975
Hemiteles pedestris (Fab.)	1960/3	**SM**	BH/AS	1975
Ichneumon extensorius L.	1960/3	**SM**	BH/AS	1975
Lissonota bellator (Grav.)	1960/3	**SM**	BH/AS	1975
Phaeogenes fuscicornis Wesm.	1960/3	**SM**	BH/AS	1975
P. planifrons Wesm.	1960/3	**SM**	BH/AS	1975
Pimpla turionellae (L.)	1960/3	**SM**	BH/AS	1975
Zootrephus rufiventris Thom.	1960/3	**SM**	BH/AS	1975

Eurytomidae

Eurytoma tibialis (Boh.)	1960/3	**SM**	BH/G	1975
Harmolita lineare Walk.	1960/3	**SM**	BH/G	1975

Encyrtidae

Copidosoma filicorne (Dal.)	1960/3	**SM**	BH/G	1975
Encyrtus dius (Walk.)	1960/3	**SM**	BH/G	1975
Microterys tessellatus (Dal.)	1960/3	**SM**	BH/G	1975

Lamprotatidae

Dicyclus circulus Walk.	1960/3	**SM**	BH/G	1975

Pteromalidae

Asaphes vulgaris Walk.	1960/3	**SM**	BH/G	1975
Cyrtogaster vulgaris Walk.	1960/3	**SM**	BH/G	1975
Meraporus graminicola (Walk.)	1960/3	**SM**	BH/G	1975
Mesopolobus incultum (Walk.)	1960/3	**SM**	BH/G	1975

Eulophidae

Chrysocharis pubicornis (Zett.)	1960/3	**SM**	BH/G	1975
Diglyphus isaca (Walk.)	1960/3	**SM**	BH/G	1975
Entedon diotimis Walk.	1960/3	**SM**	BH/G	1975
Euderis viridis Thom.	1960/3	**SM**	BH/G	1975
Hemiptarsenus unguicellus (Zett.)	1960/3	**SM**	BH/G	1975
Pediobius epigonus (Walk.)	1960/3	**SM**	BH/G	1975
Tetrastichus conon Walk.	1960/3	**SM**	BH/G	1975

Diptera Nematocera

Tipulidae

Dicranomyia sera (Walk.)	1960/3	**SM**	BH	1975
Erioptera stictica (Mg,)	1960/3	**SM**	BH	1975

Tipula paludosa Mg.	1960/3	**SM**	BH	1975

Ceratopogonidae

Dasyhelea flavoscutellata (Zett.)	1960/3	**SM**	BH	1975

Chironomidae

Halocladius varians Staeg.	1960/3	**SM**	BH	1975
Thalassosmittia thalassophilus (Goetgh.)	1960/3	**SM**	BH	1975

Sciaridae

Schwenckfeldina carbonaria (Mg.)		**SM**	BH	1975

Bibionidae

Bibio johannis (L.)	1960/3	**SM**	BH	1975
B. marci (L.)	2. V. 76	**DS**	MS	(n.r.)
Dilophus febrilis L.	2.V. 76	**DS**	MS	(n.r.)
D. femorata (Mg.)	1960/3	**SM**	BH	1975

Diptera Brachycera

Stratiomyiidae

Nemotelus notatus Zett.	12. VII. 75	**DS**	BH/MS	(n.r.)
N. uliginosus (L.)	1960/3	**SM**	BH	1975

Asilidae

Philonicus albiceps (Mg.)	17. VII. 71	**AD**	PJC	1975

Therevidae

Thereva annulata Fab.	17. VII. 71		PJC	1975
T. bipunctata Mg.	17. VII. 71		PJC	1975
T. nobilitata Fab.	17. VII. 71		PJC	1975

Bombylidae

Phthiria pulicaria Mikan	24. VI. 43	V	AWS	1975

Empididae

Drapetis curvipes (Mg.)	1960/3	**SM**	BH/KS	1975
Empis femorata Fab.	1960/3	**SM**	BH/KS	1975
Rhamphomyia tarsata Mg.	1960/3	**SM**	BH/KS	1975
Platypalpus notatus (Mg.)	1960/3	**SM**	BH/KS	1975
P. pallidiventris (Mg.)	1960/3	**SM**	BH/KS	1975

Dolichopodidae

Campsicnemus armatus (Zett.)	1960/3	**SM**	BH	1975
Dolichopus brevipennis Mg.	1960/3	**SM**	BH/O	1975
D. diadema (Hal.)	1960/3	**SM**	BH	1975
D. nubilis Mg.	1960/3	**SM**	BH	1975
D. plumipes (Scop.)	1960/3	**SM**	BH	1975
D. urbanus Mg.	1960/3	**SM**	BH	1975
Hydrophorus oceanus (Macqt.)	1960/3	**SM**	BH	1975
Machaerium maritime Hal.	1960/3	**SM**	BH	1975
Sympycnus desoutteri Parent	1960/3	**SM**	BH/O	1975
Syntormon pallipes (Fab.)	1960/3	**SM**	BH/D	1975

Diptera Cyclorrhapha

Syrphidae

Episyrphus balteatus (Deg.)	7. IX. 72	**DA**	MS	1975
Eristalis abusivus Coll.	7. IX. 72	**DG**	MS	1975
E. arbustorum (L.)	7. IX. 72	**DG**	MS	1975
E. intricarius (L.)	27. VII. 72	**DG**	MS	1975
E. tenax (L.)	7. IX. 72	**DG**	MS	1975
Lejogaster metallina (Fab.)	7. IX. 72	**DG**	MS	1975
Melanostoma mellinum (L.)	7. IX. 72	**DG**	MS	1975
Metasyrphus corollae (Fab.)	7. IX. 72	**DG**	MS	1975
M. latifasciatus (Macqt.)	27. VII. 72	**DG**	MS	1975
M. luniger (Mg.)	7. IX. 72	**DG**	MS	1975
Platycheirus albimanus (Fab.)	7. IX. 72	**DG**	MS	(n.r.)
P. angustatus (Zett.)	7. IX. 72	**DG**	MS	1975
P. clypeatus (Mg.)	27. V. 71	**DG**	MS	1975
P. manicatus (Mg.)	25. IV. 71	**DS**	MS	1975
P. scutatus (Mg.)	27. VII. 72	**DG**	MS	1975

Pyrophaena granditarsa (Forst.)	1960/3	**SM**	BH	1975
Scaeva pyrastri (L.)	1960/3	**SM**	BH	1975
Sphaerophoria menthastri (L.)	30. VIII. 72	**DG**	MS	1975
S. philanthus (Mg.)	2. V. 76	**DS**	MS	(n.r.)
S. rueppellii Wied.	12. VII. 75	**DS**	BH/MS	1976
S. scripta (L.)	27. VII. 72	**AD**	MS	1975
Syritta pipiens (L.)	7. IX. 72	**DG**	MS	1975
Syrphus ribesii (L.)	27. VII. 72	**DG**	MS	1975
S. vitripennis Mg.	7. IX. 72	**DS**	MS	1975
Volucella bombylans (L.)	27. VII. 72	**DG**	MS	1975

Tephritidae

Paroxyna plantaginis (Hal.)	1960/3	**SM**	BH	1975

Otitidae

Herina frondescentiae (L.)	12. VII. 75	**DS**	BH/MS	(n.r.)

Psilidae

Psila nigra Fal.	27. V. 71	**DG**	MS	(n.r.)

Helcomyzidae

Heterocheila buccata (Fal.)	26. VI. 76	**DS**	MS	(n.r.)

Chamaemyiidae

Chamaemyia herbarum (R.–D.)	1960/3	**SM**	BH/KS	1975

Coelopidae

Coelopa frigida (Fab.)	1960/3	**SM**	BH	1975

Heleomyzidae

Heteromyza commixta Coll.	26. VI. 76	**DS**	MS	(n.r.)
Morpholeria kerteszi (Cz.)	1960/3	**SM**	BH	1975

Sepsidae

Sepsis violacea Mg.	1960/3	**SM**	BH	1975

Sciomyzidae

Limnia unguicornis (Scop.)	26. VI. 76	**DS**	MS	(n.r.)
Pherbellia cinerella Fal.	14. II 71	**AD**	MS	(n.r.)
Salticella fasciata (Mg.)	1960/3	**SM**	BH	1975

Sphaeroceridae

Copromyza uncinata (Duda)	1960/3	**SM**	BH/D	1975

Piophilidae

Piophila nigricornis Mg.	2.V. 76	**DS**	MS	(n.r.)

Ephydridae

Glenanthe ripicola Hal.	1960/3	**SM**	BH/D	1975
Hyadina humeralis Beck.	1960/3	**SM**	BH/D	1975

Tethinidae

Tethina grisea (Fal.)	1960/3	**SM**	BH/D	1975
T. illota Hal.	1960/3	**SM**	BH/D	1975

Agromyzidae

Phytomyza varipes Macqt.	1948	**DG**	G	1968

Chloropidae

Chloropisca glabra (Mg.)	1960/3	**SM**	BH	1975
Chlorops troglodytes (Zett.)	1960/3	**SM**	BH	1975
Melanum lateralis Hal.	1960/3	**SM**	BH	1975
Oscinella frit (L.)	1960/3	**SM**	BH/Cr	1975
O. pusilla (Mg.)	1960/3	**SM**	BH/D	1975
O. vastator Curt.	1960/3	**SM**	BH/D	1975
Thaumatomyia notata (Mg.)	1960/3	**SM**	BH/D	1975
Tricimba cincta (Mg.)	1960/3	**SM**	BH/D	1975

Scathophagidae

Scathophaga litorea (Fal.)	1960/3	**SM**	BH/KS	1975
S. stercoraria (L.)	1960/3	**SM**	BH	1975

Calliphoridae

Calliphora vicina R.–D.	1960/3	**SM**	BH/Cr	1975

Muscidae

Coenosia albatella (Zett.)	1908	**U**	H/C	in pr.
C. antennata (Zett.)	1960/3	**SM**	BH/KS	1975
C. dubiosa Hennig	12. VII. 71	**DG**	C	1975
C. femoralis R.-D.	1960/3	**SM**	BH/KS	1975
C. pedella (Fal.)	1960/3	**SM**	BH/KS	1975
C. pulicaria (Zett.)	1960/3	**SM**	BH/KS	1975
C. tigrina Fab.	1960/3	**SM**	BH/KS	1975
C. verralli Coll.	1960/3	**SM**	BH/KS	1975
Dexiopsis lacustris Karl	12. VII. 71	**U**	C	in pr.
D. litoralis (Zett.)	1908	**U**	H/C	in pr.
D. minutalis (Zett.)	1960/3	**SM**	BH/KS	1975
Phaonia incana (Wied.)	1960/3	**SM**	BH/KS	1975

Siphonaptera

Pulicidae				
Pulex irritans L.	?	**U**	O'M	1966
Spilopsyllus cuniculi (Dale)	?	**U**	O'M	1966
Hystrichopsyllidae				
Ctenophthalmus nobilis (Rothsch.)	?	**U**	O'M	1966
Typhloceras poppei Wag.	?	**U**	S	1966
Ceratophyllidae				
Ceratophyllus garei Rothsch.	?	**U**	O'M	1966
Nosopsyllus fasciatus (Bosc)	?	**U**	O'M	1966

INDEX

(Italic page numbers refer to colour plates.)